The Changing Era of Diseases

The Changing Era of Diseases

Yun-Chul Hong

ACADEMIC PRESS

An imprint of Elsevier

Academic Press is an imprint of Elsevier
125 London Wall, London EC2Y 5AS, United Kingdom
525 B Street, Suite 1650, San Diego, CA 92101, United States
50 Hampshire Street, 5th Floor, Cambridge, MA 02139, United States
The Boulevard, Langford Lane, Kidlington, Oxford OX5 1GB, United Kingdom

Library of Congress Cataloging-in-Publication Data
A catalog record for this book is available from the Library of Congress

British Library Cataloguing-in-Publication Data
A catalogue record for this book is available from the British Library

ISBN: 978-0-12-816439-6

For information on all Academic Press publications visit our website at
https://www.elsevier.com/books-and-journals

Working together
to grow libraries in
developing countries

www.elsevier.com • www.bookaid.org

Publisher: Mica Haley
Acquisition Editor: Rafael E. Teixeira
Editorial Project Manager: Jaclyn A Trusdell
Production Project Manager: Debasish Ghosh
Cover Designer: Matthew Limbert

Typeset by TNQ Technologies

Contents

Preface

Nearly everyone has probably been worried about the prospects of living and dying at least once in his or her lifetime. I was no exception. I have also pondered the meaning of life, death, and the passage of time and had a deep desire to understand how the society to which I belong was formed and where it is headed. In particular, one specific question has dogged me: Is it really possible to understand the history of an individual's birth, illness, and death separately from human civilization or the history of humankind? Now, I think an individual's life and eventual death can be understood correctly only when we comprehend the historical transition from the beginning of humankind to the extinction of humankind, or the emergence of the next hominid species.

This book intends to cover the history of humankind in a consolidated framework that integrates the historical development of social, political, and cultural events and biological development as evolutionary results of the physical and mental structure and function. These two both deal with human history. However, when you look at them closely, you might see that they consist of completely different ideas and approaches.

When I look back at my life, I realize I was quite fortunate to have opportunities to learn these two different approaches toward human history. The opportunity to obtain a view of history from the perspective of the humanities and social sciences came when, as a student at a university, I started worrying about how society could survive the dark ages in which freedom was suppressed and there was no democracy. Meanwhile, I was also exposed to a biological and evolutionary view of human history while studying in medical school and reading books about Charles Darwin's theory of natural selection.

In 1979, when I entered university with a fantastic dream just like any freshman, I immediately had to adjust to life in the campus, which was then overwhelmed by the atmosphere of dictatorship. It was the year when freedom was suppressed at its peak in South Korea. At the same time, it was also the year when the long-time dictator Park Chung Hee was assassinated. Before entering the university, I thought I would be enjoying college life by learning to play the violin and reading as many books as possible. However, I could not stop thinking about what was happening on campus, which was characterized by tear gas, slogans, shouting, and tears.

Two years later, a new dictator came to power. The military took complete control of the regime and went on to rule the country with an iron fist. I was then studying at the College of Medicine. In the spring, magnolias and lilacs were wonderful to look at outside the classroom, but there was no real spring in our society, which crumbled under the military dictatorship. Nevertheless, there was nothing I could do but read books to answer the question to myself, "Why is this happening in my life and to our society?" Meanwhile, I could not fully concentrate in medical school although I spent much time studying. Rather, I felt bored memorizing countless medical terms and studying each illness without clear logic. Still, books about history and philosophy kept me engaged. I was happy studying philosophy, economics, and sociology to understand the reality in our society.

Through such learning opportunities during that period, terms such as "society," "politics," "disease," and "death," which I encountered in my studies, were initially confusing because I could not grasp their clear meanings. Gradually, I began to connect them with human history or the times in which they have been significant. I must confess that the books of Karl Marx and his colleagues opened my eyes to history revealing human life. On the other hand, however, I could not abandon the suspicion that there is excessive structural scheming in Marx's argument that cannot fully explain reality. Meanwhile, Darwin, whose works I read at the end of my course in medical school, was another person who opened my eyes to the history of humankind. Although Darwinists' books faithfully emphasized Darwin's theory of evolution, I was also disappointed by their simplistic logic of natural selection. I thought neither reality nor the history of the past could be well explained by Marx or Darwin.

Even if their arguments are different from my ideas, I am grateful to them because the subject or logic developed in this book received much inspiration from Marx and Darwin. Particularly, the ideas of streaming human history were borrowed from the logic of Marx and Darwin. However, although both Marx and Darwin contributed significantly to the societal and biological inquiry domains, they could not depart from the mechanistic epistemology underlying the scientific system in 19th-century Europe, which, I believe, hindered them in fully understanding the history of humankind.

In this sense, G.W.F. Hegel's dialectics and oriental tradition of thinking helped. Without the integrated view of the world, I could not have written this book with a holistic epistemology that was based on harmony and balance. Meanwhile, Jared Diamond, whose books I read when I became a professor of medicine after finishing my course in family medicine and preventive medicine, provided me another perspective and methodology in interpreting the world. It would be fair to say that his book *Guns, Germs, and Steel* has vastly influenced me, just like Marx or Darwin did. The method of comparing and interpreting phenomena based on local and historical contexts rather than explaining history through a dogma was a fresh approach.

This book deals with the evolution of disease, our response, and the view of the future society. I would be too ambitious if I included all these major topics in one book, so I focused on describing how each theme and corresponding content are connected with the development of diseases in one entire context, rather than discussing each one in painstaking detail.

In a certain aspect, the contents of the book can be described logically by understanding the world through rational epistemology, but to describe true human history, one must directly experience the core parts of human affairs, such as love, conflict, disease, marriage, and family. Still, it is difficult to reflect on the meaning of the terms exactly as we experience human affairs in our lives. Therefore, I am deeply grateful to those who have experienced the joys and the sorrows of life with me, providing me an understanding of our lives. Without such experience of human life, this book would not have been completed.

I sincerely thank all my colleagues and students for their assistance. Specifically, it would not have been published without the help of Seonhee Kwon, Nami Lee, and Hisashi Ogawa, who lent worthy advice on the contents and the flow of the book. Above all, I thank my wife, Eunhee Ha, and our two daughters for their support and patience while I was writing this book. I realized that family is the most important unit of the *Homo sapiens* community, and the meaning of family became even deeper throughout the course of writing this book.

Overview

1. Can We Live in a Disease-Free World?

Having a disease-free world is everyone's dream, but such a dream will come true only through concerted efforts in various sectors, including medical science, engineering technology, economy, and politics, which are associated with the resolution of diseases. Probably, the precondition for the success of such efforts is sharing a sound perspective on health and disease. Health can be understood as a state of life that allows not only the current generation but the next generations to survive in harmony with the surrounding environment. As the environment changes constantly, however, the relationship between humans and the environment cannot be kept in a state of harmony and balance, and if the scale is tipped, a crack will inevitably occur. Disease occurs when this crack cannot be filled. Therefore, knowing which changes in the environment lead to disease and understanding the future diseases as well as the current diseases of humankind is the first step toward conquering disease.

In this regard, it is necessary to understand the relationship between the changing environment and the diseases caused by such environmental changes from the history of humankind. It is especially important to look at the history of humankind after the advent of civilization, during which the living environment of humans changed greatly. Disease was born in earnest with the dawn of civilization, which was prompted by the agricultural revolution that occurred some 10,000 years ago. Disease has since emerged as one of the critical factors threatening human lives, along with war and famine. In response, humans have exerted great efforts to understand and overcome disease. Therefore, it is necessary to understand the development of medical concepts and practices along with the historical development of the community that makes up the living environment of humans as well as the politics, economics, culture, and philosophy surrounding it. In fact, one cannot understand diseases properly from the biological or technological perspective alone. Not until we expand our perspectives beyond individuals and consider the life and death of individuals within the community and the human species can we understand disease correctly.

In the early days of the Agricultural Revolution, the people could not obtain enough food due to low productivity and the resulting insufficient agricultural products. As such, people suffered from diseases arising from malnutrition as they could not intake sufficient nutrients due to their limited crops. Moreover, the infectious diseases transmitted by livestock and the community life that easily propagated them threatened human health, so humans did not live for more than 30 years on average. As the productivity increased and the size of the community grew, however, human life more or less stabilized and a new political power emerged in the form of the nation state. As the state by nature wanted to gain more profit by exploiting or using other areas, regional or interregional wars broke out more frequently, while also promoting trades and exchanges. Such changes brought about new types of diseases that had not been experienced by humans before, paving the way for the "Age of Epidemics." Human contact with new germs led to the outbreak of serious epidemic diseases among the people who either failed to acquire immunity or did not enjoy a sufficient adaptation period.

Now, hundreds of years have passed since the Industrial Revolution in the 18th century, which brought about remarkable advancement in productivity. This created an entirely new relationship in terms of production, where the capitalists pursued profit and the workers sold their labor for living. In the early days of the Industrial Revolution, the living conditions of the working class were unfavorable, causing many diseases. As the living environment improved along with scientific development and productivity improvement later, the working class, which was composed of the general public, enjoyed a materially abundant life as well, greatly reducing the infectious disease epidemics and thereby ending the Age of Epidemics. However, the explosive growth of many different chronic diseases transpired, such as diabetes mellitus, obesity, hypertension, and cancer, which are collectively known as chronic diseases, the diseases of modern humans. In short, the Industrial Revolution became a hotbed of all sorts of chronic diseases that continue to dog humans today.

2. Pandemic of Chronic Diseases and the Late Chronic Diseases

Until the early 20th century, pneumonia, tuberculosis, and gastroenteritis were the major causes of death worldwide, accounting for one-third of all deaths. According to a report by the World Health Organization, however, 68% of all human deaths in 2012 were due to chronic diseases such as heart disease, diabetes mellitus, and cancer. Not only have the major diseases responsible for death changed within just a century; the proportion of deaths due to chronic diseases also increased greatly among all human deaths.

Unlike infectious diseases, chronic diseases are not caused by a single pathogen but have multiple causes. Not only does it take a considerable amount of time for such diseases to develop; the patients also continue to suffer from such diseases for an extended time, not dying immediately or recovering quickly. Whereas infectious diseases are "old" diseases caused by germs, chronic diseases are caused by the inability of the human genes and the living environment to harmonize with and adapt to each other. The reason is that the modern humans with genes adapted only to the past environment are exposed to various new environments they have never experienced before. It takes a considerable amount of time for the human genes to change and to adapt to their new environment, and as the rate of change in the contemporary society is much faster than the speed of genetic adaptation, humans are now experiencing a pandemic of chronic diseases due to such incongruity. In other words, the rapid changes in the environment are among the major causes of diseases in the 21st century.

Compared with the hunter-gatherers, modern humans have greatly increased calorie intake due to the surplus food, and the composition of their food has changed much as well, with excessive intake of sugar, salt, and animal fat along with the insufficient intake of vegetables. Moreover, the amount of physical activity of modern humans has been greatly reduced, and people have picked up new harmful habits, such as alcohol drinking and cigarette smoking. Modern humans are also exposed to new chemicals, such as air pollutants and endocrine disruptors, while living in a more competitive society. Due to such changes in modern humans' living environment, the genes that had been beneficial to survival in the past are now more likely to cause disease. This is the fundamental reason for the explosive spread of chronic diseases, such as diabetes mellitus, hypertension, heart disease, and cancer.

What is interesting here is that the prevalence of chronic diseases is also changing rapidly. It has already been shown that not only the deaths due to cardiovascular disease and cancer are declining in developed countries after reaching the peak levels; the incidence rates of such diseases are also declining. As the incidence rates of chronic diseases in developing countries are currently increasing, like epidemics, the incidence rates of such diseases are still growing for the entire human population but will be reduced before long. If the increasing incidence rates of chronic diseases start to show a downward trend, can humankind be said to be truly entering a "disease-free age"? Even if the incidence rates of chronic diseases are successfully reduced, it is most likely that humankind will still face new problems. Moreover, in the same way that the chronic diseases exploded after the contraction of infectious disease epidemics, new diseases can grow rapidly as the chronic diseases are being reduced.

The newly emerging diseases include neurodegenerative diseases such as Alzheimer disease and Parkinson disease, immune system disorders such as atopy and Crohn disease caused by disturbed immune function, and mental disorders triggered by increasingly excessive mental exhaustion such as competition and stress. These diseases are usually chronic or progressive and are caused by multiple factors but are also characterized by new additional factors, such as aging, disturbed intestinal microbiota, and the competitive social structure, in addition, to various existing factors known to trigger chronic diseases. Therefore, although they are chronic diseases, they can be named "late chronic diseases" to distinguish them from the existing chronic diseases.

Meanwhile, just as urbanization provided a hotbed of chronic diseases, globalization can trigger another epidemic of infectious and environmental diseases. As a result of globalization, there have been new epidemics of viral infections, such as severe acute respiratory syndrome, Middle East respiratory syndrome, and Zika virus. Moreover, if a new pathogen, such as avian influenza virus, infects humans by taking them as hosts, a pandemic spread of a new infectious disease can occur. The meltdown of the Fukushima nuclear power plant in Japan and the serious air pollution in China indicate that environmental diseases can also spread across borders or regional boundaries.

3. Changing View of Disease

Just as illness has changed over time, humanity's view of illness has also undergone historical changes. As the ancient civilizations like that of Athens were built, humanity began to look at the world with rational eyes, contrary to the previous religious viewpoint that considered human disease and death punishments "inevitability ordained by the gods." They have brought the disease from the "divine domain" to the "human domain." The epidemic spread of infectious diseases, however, which began near the fall of the Roman Empire, revived the view of disease as something belonging to the divine domain. As there was no proper way to deal with the epidemic that spread at that time, the past view of disease, that it is a punishment by a bad spirit or a god, reappeared. The idea of disease came into the realm of science in earnest only after the Renaissance and the Industrial Revolution, which took place more than 1000 years after the reappearance of such a view of disease.

Since then, the "biomedical view of disease," which was established by looking at diseases from a scientific perspective, has been deeply rooted as the central idea of modern medicine, exerting much influence to this day. The biomedical view of disease is based on the idea that external elements such as bacteria invade the human body and impair the structure or function of a cell in a specific organ, causing disease. In the biomedical view of diseases, biologic elements such as specific microorganisms or genes are considered directly causing diseases, but such a view of pathogenesis has lost its explanatory power later because there are few environmental factors or genetic variations that can solely explain the development of chronic diseases. On the contrary, it has been discovered that chronic diseases occur as endogenous systems of the human body, such as genes, proteins, and metabolites, get disturbed with entanglement with numerous environmental factors.

In fact, diverse genes are involved in a single disease, and there are also many cases where a certain gene is associated with multiple chronic diseases. As environmental change also takes diverse forms depending on the individual or population group, it is difficult to simplify disease risk factors. It is also difficult to isolate the independent effect of a specific factor in the development of a disease as the changes in the main components of life (e.g., food, physical activity, and human relationships), pathogenic germs, and the environmental threats influence each other and team up in triggering chronic diseases. In the case of diabetes mellitus, hypertension, arteriosclerosis, hyperlipidemia, and heart disease, each of them is regarded as a different disease from the others, but it would be more reasonable to assume that such diseases do not occur independently but in an interrelated fashion because more than one of such diseases often occurs in an individual.

The phenomenon in which a specific factor causes a specific disease in a simple one-to-one relationship between the cause of a disease and its pathological outcome is rarely seen in chronic diseases. Therefore, the efforts to isolate a single factor as a prime cause of disease are not valid in explaining most chronic diseases because the disease is caused by a variety of internal and external factors of the human body as they push the structure and functions of the human body to deviate from the normal ranges. Therefore, it is necessary to establish a new medical model beyond the conventional biomedical view of pathogenesis: the systems medicine model, which is based on the idea that a disease occurs when the balance and harmony of the multidimensional systems inside and outside the human body is broken.

Along with the changes in the conventional view of disease, patient care in hospitals must change as well because the current disease-centered care practices have exposed a number of serious problems. Although disease-centered medical care has contributed greatly to the enhancement of medical expertise, it is difficult to treat patients efficiently with the concept of responding to each disease independently. In the end, the medical practices must be changed from disease-centered medicine to people- or patient-centered medicine, with the corresponding changes in the medical education. The future physicians should acquire the knowledge and skills to understand the various pathological phenomena in the relationship between the internal system of human body and the external environment and to use an integrated and holistic approach in treating their patients.

In the future, comprehensive health management incorporating methods for overcoming physical, psychological, and social dysfunction should be conducted based on a holistic understanding of the individual patients' situations rather than just concentrating on individual diseases. As diagnoses and prescriptions become more precise, and as cell and tissue regeneration as well as gene and human function strengthening is practiced more actively in the future, individualized or precision medical care will be used to deal with individual health problems. In such an age, traditional medical practices, such as disease diagnosis and treatment in the hospital, will be only a part of the comprehensive medical care. In other words, humanity will see an age of new disease management where various lifestyle and environmental factors that cause disease or affect disease progression are identified, and continuous and comprehensive healthcare services are provided to patients to correct them.

4. What Next After the End of "the Era of Diseases"?

When viewed from the perspective of the future generations, this era may be remembered as an age of the most remarkable changes in the disease patterns and the most significant increase in biological longevity. Over the past 150 years, the average life expectancy of humankind has jumped nearly two- or threefold. For instance, the average lifespan of people has remarkably increased by 6 months almost every year over the past 60 years in South Korea. It is only the human race among all the species that has achieved a several-fold lifespan increase within such a short period. Such an increase in the lifespan of humankind is basically due to the reduced mortality from disease. Can humans, however, celebrate the utopian world they have been dreaming of if the developments in science and medical technology push humanity to enter an era where disease is completely controlled and humans can survive beyond the limits of their biological life?

Even if a technology arises that can treat chronic or even late chronic diseases as well as most infectious diseases, it will not easily mark the end of human disease and the dawn of a blissful future for humanity. If the present economic inequality, unbalanced development of science and technology, and differences in medical accessibility will be sustained or even intensified in the future, there will surely be those who will not benefit from the developments in medical technology, and who will still suffer from illness. On the contrary, there will be some who will asymmetrically enjoy the achievements of science and technology. In the end, only the latter will possess outstanding physical capabilities through the strengthening of their biological functions. As soon as such "biological inequality" becomes a reality, the future society will cross the Rubicon, entering an age of conflict and confrontation with no hope of reconciliation.

Therefore, the future society can pose to humankind more serious challenges than infectious disease epidemics or chronic diseases. Moreover, the problems that humans may face in the future are not only those that they have not yet experienced but also those that cannot be easily solved, such as population aging, the choice and management of death, and a stagnant society that has lost its vitality. Ultimately, humankind will hold the key to solving their difficult problems, such as chronic illness and aging, in the future, but such key may also be one that will open the door to new challenges.

Now humankind stands at a critical juncture of destiny. Our fate, or whether we will have a bright future liberated from disease or will face serious survival challenges owing to the changes our society will undergo, will depend on how we respond to the challenges. Therefore, efforts must be made not only to conquer disease but also to ensure that the changes that will be triggered by the control of diseases will not give rise to another crisis of humanity. Now is the right time to start using personal strategies to control disease and, at the same time, to implement social and global strategies at the community level as well, along with efforts to come up with a medical system that can cope with the future changes appropriately.

5. Composition of This Book

The most important purpose of this book is to methodologically explore how to end humanity's suffering from illness, but at the same time, it attempts to explain the challenging problems that may arise from the end of disease in the future. With such perspectives, this book reviews the changes, conquest, and future of disease, with focus on the following issues.

The first issue that will be addressed by this book is that disease is in line with historical changes. The transition of disease from that in the era of hunter-gatherers to chronic diseases in the modern society indicates that disease occurrence is closely linked with the socioeconomic conditions of humans. Therefore, the advancement of productivity propelled by automation and computerization, the corresponding changes in the production relations, and the political and economic changes in the wake of globalization will also affect the future disease patterns. These changes will not only give rise to changes in the community, such as urbanization and globalization, but will also push humanity to the era of new emerging diseases such as late chronic diseases, with the frequent occurrence of degenerative diseases, immune disorders, and mental illnesses.

Second, the concept of mechanistic causality, in which disease is regarded as a phenomenon caused by an abnormality in a specific cell, tissue, or organ of the human body, does not allow us to properly understand chronic diseases and bring about the eventual control of diseases. Therefore, disease awareness should develop from the biomedical perspective of disease to a complex and systemic perspective of disease. Based on this awareness, efforts to regain the balance and harmony between the internal and external systems of the human body are required to conquer diseases.

Third, the development of science and technology will not only increase the overall health level of humans but also prolong the human lifespan by reducing the obstacles to the extension of the biological lifespan of humans. On the other hand, however, when this becomes a reality, the elderly will comprise the majority of the population because not only people live longer, but the motivations to give birth to offspring will decrease, along with the birth rate. In such society, humans may face new challenges, ones that they have never experienced before.

This book consists of five parts. In Chapter 1, I explain why the history of disease should be seen alongside the development stages of civilization, while reviewing respective periods from the beginning of diseases until the pandemic spread of infectious diseases. Chapter 2 explains how chronic and late chronic diseases emerged, how disease patterns can be expected to change in the future, and how "systems medicine" can help solve such diseases. In Chapter 3, the five strategies used by our bodies to conquer diseases are discussed: (1) the symbiotic system, which cooperates with microorganisms; (2) the poison metabolism system, which strengthens our bodies' defense against toxic substances; (3) the immune system, which protects itself from external intruders; (4) the healthy aging system, which ensures a healthy aging process; and (5) the regeneration system, which strengthens our body functions. In addition, how the aforementioned systems work in our bodies is explained, and strategies to deal with them are also explored. In Chapter 4, the changes in the healthcare system as a response to the changing diseases are looked into, and individual practices as well as social and global strategies that could end disease are explored. Last, in Chapter 5, the limitations of the human lifespan and the impact of the changing lifespan on families and social communities are discussed. Finally, the prospects of the Anthropocene epoch, in which humanity is exerting a profound impact on the entire global environment, are presented.

In conclusion, this book looks at the changes in disease and medicine from a historical perspective and explains how humanity can conquer disease, a long-cherished desire that humankind has been dreaming of. Sooner or later, the time when diseases have had a profound impact on human life, death, and longevity will fade away. Explaining such perspectives, I hope that this book will be read as a guide for the conquest of diseases and as a prospect for the future.

Chapter 1

From the Origin of Disease to Pandemic Infectious Diseases

1.1 The Emergence of Diseases

As the climate became warmer in the mid-latitudes in the wake of the conclusion of the last ice age, humankind entered a civilized society based on farming and herding. The first civilization grew as farming began in the Fertile Crescent located in the east of the Mediterranean Sea. Then, civilization spread mainly to the regions in similar latitudes where the climatic conditions were relatively comparable, thereby prompting the development of civilizations along the east–west axis. The supply of nutrients to humanity greatly changed compared with those supplied in the age of the hunter-gatherers, exerting a considerable influence on the disease pattern. In addition, animal domestication provided a hotbed for various infectious diseases.

Humankind undergoes physical and intellectual development

From the time of the prehistoric Australopithecus to the time of the *Homo sapiens*, the culture of the hunter-gatherers showed that incremental changes and developments had been made in stoneware, fire, food, and communication. Stoneware was a very important cultural tool that gradually evolved from sharp crudely carved stone pieces to finely polished axes and knives. By about 1.2 million years ago, our hominid ancestors became capable of dealing with tools better than before. This enabled them to hunt and eat animals using hunting tools as well as to cook meat safely using fire. Animal hunting and meat eating accelerated brain and body development, laying the foundations for the emergence of humans.[1]

It was about 50,000 years ago that humankind began to fish and cobble clothes together using more sophisticated tools. By this time, humans had become more omnivorous, eating increasingly diverse foods, including the meat of larger animals and diverse aquatic products. They were also more capable of surviving in cold climates as they began to wear clothes. In other words, humans had become equipped with the basic tools that would enable them to go out of Africa and create a culture in earnest. Increasingly elaborate stoneware continued to appear, and ceremonial cultures emerged, along with cave paintings, sculptures, and ornaments.

The Changing Era of Diseases. https://doi.org/10.1016/B978-0-12-816439-6.00001-6

These cultures were not created simply because people were equipped with technology to fabricate but because they could engage in language-based communication. Therefore, it can be assumed that humankind was already equipped at that time with the biological capability of creating a culture based on communication. For instance, a brain with considerable intelligence, the physical structure of the larynx, and vocal cords for vocalization as well as the characteristic oral movement for speaking are required for humans to speak languages. It means, therefore, that humankind had the physical and intellectual capabilities to create cultures at this time. Unlike other predators, humans are equipped with an excellent ability to hunt the prey and gather foods through communication-based collaboration while lacking the excellent bodily structures required for hunting, such as claws and sharp teeth. Given that there is an opportunity to produce more offspring by mating if food can be obtained relatively easily, the ability to collaborate based on communication using language can be assumed to have been the main characteristics that humans acquired under the pressure of natural selection.

Our hunter-gatherer ancestors often left the places where they stayed to find new settlements when the population grew overly or the surrounding environment changed significantly, or when there was no longer enough food to feed them. At the beginning, when they were in Africa, they would have formed small groups, composed mainly of their family members, and moved around frequently. At first, they secured their own territories at a distance from other groups, but the groups needed to come in contact with the other groups to create new families by mating, and they gradually formed a community beyond the barriers of their families by repeating such process. About 50,000 years ago, they left Africa and spread to the four corners of the globe, while adapting to different regions, branching into different races, and diversifying their genetic characteristics.

The glaciers covering the mid-latitude region withdrew, exposing grasslands, rivers, and lakes, as the last ice age, which began 30,000 years ago, ended 12,000 years ago. Moreover, the temperatures then, which were some 5–6°C higher than those in the ice age, changed the earth's vegetation and fauna, making them more diverse. The conditions conducive for agriculture, which was the basis of civilization, thus began to form. At this time, the intellectual power of mankind further improved, and flashes of human creativity began to appear more clearly. Tools such as stoneware were further developed while cave paintings improved human artistry. Also, thanks to the warmer climate, the earth's ecosystems became more diverse, providing a relatively rich amount of food for mankind, although it fluctuated depending on the seasonal or environmental conditions. Humankind's new diet consisting of various food items, such as animals, plants, fruits, nuts, fishes, and shellfishes, stimulated the physical and intellectual development of humans more. Through such changes, mankind was able to establish cultural norms and the biological bases for the future civilizations.

Hunter-gatherers suffered from infectious diseases but never from communicable diseases

The power of nature was still more dominant than the power of cultural tools such as stoneware and language in the age of the hunter-gatherers as civilization had yet to be formed. With the transition from the hominid ancestors to the *H. sapiens*, the conditions for survival gradually improved, along with the use of tools and fire and changes in food, clothing, and settlements. The mortality rate, however, remained almost the same as the birth rate, and there was little increase in population, with the average human lifespan being 20–25 years and with hardly any person living beyond 40 years. The main health threats posed to the hunter-gatherers were lack of food and injuries acquired while engaging in hunting. The bacteria in the living environment could infect their wounds, which posed threats to their life. The vast majority of the communicable diseases seen today, however, rarely broke out during the age of hunter-gatherers.

The bacteria that cause infectious diseases are only a few among the countless microorganisms that are found inside and outside the human body and in the living environment. In fact, the vast majority of microorganisms do not cause disease but help humans by forming symbiotic relationships with them. Most of such symbiotic relationships, however, occur only when humans and microorganisms maintain a balance of power between them. When the nutritional status of a person is poor or when he is under great stress or suffers immune degradation, the balance of power between the two parties is broken, prompting the microorganisms that are in a symbiotic relationship with the person to turn to cause a disease. For instance, *Escherichia coli*, a bacterium that normally lives in the large intestine, does not usually cause disease but can infect its host if the human immune system fails or if the colon environment changes significantly.

Specifically, if a person has a high degree of defense capability and the power balance between him or her and the microorganisms inside his or her body is upheld, the symbiotic relationship between the person and the microorganisms can be sustained. This relationship, however, was not present when the hominid ancestors or humans first encountered microorganisms. Perhaps microbes such as amoebas and bacteria had not yet adapted to their host when they first entered the bodies of the hominid ancestors or humans. Therefore, microbes may have caused diseases in the individuals or groups, killing at least some of them. Ultimately, natural selection took over, and only those individuals who did not acquire a disease or did not become sick would survive when these microorganisms entered the body. Microorganisms have to die as well if the host infected with a disease dies. Accordingly, it is not desirable for the host to die from the standpoint of the microorganisms, particularly in terms of survival and reproduction. Therefore, even in the case of microorganisms, natural selection has occurred in a direction that causes less disease of the host. The coexistence or symbiotic relationship between the microorganisms and humans formed through such process can be understood as a balanced state between the microorganisms' toxic ability to infect and the defense capability of humans.

On the other hand, for microorganisms to cause diseases in humans and to further spread these, it is necessary to penetrate the human defenses, but there are other important preconditions as well. First, microbes must be able to reproduce and spread themselves by taking humans as hosts, but hunter-gatherers did not live in one place as a large group at the time. Instead, they lived by forming small groups, each away from the others, or moved often to various places for finding foods, making it difficult for the microorganisms to spread among people or for the intermediary hosts to transmit the germs to other people. For instance, a measles-causing virus will die when it fails to spread to other people immediately even though it infected a person successfully. Eventually, the measles would spread easily in those areas where many people lived but not in those places that were not crowded with people. Therefore, in the age of the hunter-gatherers, the infectious power of pathogenic germs would be limited because only those germs capable of maintaining the infected state for an extended period, such as Hansen bacillus, would have an opportunity to infect other people.

When people hunt animals or deal with dead meat, the germs of the hunted animal can enter such people and make them acquire an infectious disease. As humans, however, are not normal hosts of such germs in the vast majority of cases, they do not spread easily among people. Moreover, if one is exposed to the toxins of the anaerobic bacteria in the intestines or in the decayed parts of infected animals, one can acquire a life-threatening disease, such as botulism or tetanus, even though not directly infected by such bacteria. As these toxins, however, do not possess infectious power, they cannot spread to other people. In other words, in the age of the hunter-gatherers, even if the people got infected with germs or got poisoned through the animals that they hunted, the germ rarely spread among the people and thus rarely transmitted diseases, causing epidemics.

Civilization gave birth to diseases

As the environmental conditions of various regions became diverse in the wake of the conclusion of the last ice age, the living environment became diverse as well, thus differentiating the food acquisition pattern, which is closely linked with the surrounding environment. The people living in the coastal regions or near a river were able to acquire fish or shellfish, and those living on grasslands mostly hunted herbivores. The vast majority of people, however, turned to agriculture, choosing to grow grains like wheat, barley, and rice thanks to the changed environment, which was conducive to growing plants and diverse flora. Humankind finally entered the age of civilization in earnest, in which farming, livestock, and fishery became the basis of life as a result of conversion from the hunter-gatherers' lifestyle to the living of farmers and herders.

When the communities began to form civilized societies some 10,000 years ago, such communities were mainly made up of a few families, but these small communities later developed into bands, tribes, and nations as they formed alliances among themselves or competed with one another. The communities that were left out of such competition either disappeared or were forced to settle in the remote areas, including tropical rainforests, polar regions, or deserts, far away from the central community. Although community had continued to change to some degree in the hunter-gatherer period, the formation, movement, and settlement of such communities were further accelerated with the beginning of civilization. In addition, to the natural environment, such as the varying daylight hours, temperature, altitude, and precipitation, and the seasonal changes, the lifestyles spurred by civilization, such as the farming and herding lifestyles, came to exert a great influence on the disease transition.

Above all, as humankind entered the realm of a civilized society based on farming and herding, the supply of nutrients greatly changed compared with those supplied in the age of the hunter-gatherers, exerting a considerable influence on the disease pattern. While farming and herding ensured a stable supply of food, unlike hunting and food gathering, the number of available food items became much less diverse. If one settles in a certain area, without moving any further, and depends on specific harvests, it is expected that the stability of the food supply will increase but the food diversity will decrease. Even in the case of a settled life, however, it is still possible to eat as many different foods as possible when the food exchange between regions is intensified. On the other hand, the diversity of the available food items was inferior to that in the age of the hunter-gatherers in most areas that were isolated from other areas or where commerce and trade had not been fully developed.

Our hunter-gatherer ancestors were able to ingest at least 100 different foods thanks to the seasonal variety of available foods, even though they could not expect a stable food supply. In the agricultural society, however, the number of available food items was reduced to 10–15, although it was possible to eat relatively stable foods through crop production.[2] As a result of such reduced food variety, the lack of balanced nutritional intake often resulted in a deficiency of essential nutrients. For example, the intake of carbohydrates greatly increased after the introduction of agriculture, but the intake of proteins such as those from meat was significantly reduced. In the end, the diets limited to only a few nutritional sources caused previously nonexistent diseases like rickets, scurvy, and pellagra. In the case of the lifestyle based on herding or fishery, the protein intake was not lower than that of our hunter-gatherer ancestors, but the diversity of food items was also greatly reduced, often resulting in an imbalance in the body's nutritional supply.

In addition, ever since people started farming and herding, and thus settling in certain areas, the incidence of insect-mediated diseases such as malaria has increased. In the case of insects such as mosquitoes and flies, it is difficult for them to spread a disease from one person to another if people move frequently to a distance beyond the usual activity radius of the insect.[3] Therefore, diseases such as malaria and sleeping sickness, and pathogens being spread by insects, could not prevail during the age of the hunter-gatherers, where people had to move frequently to find foods. These insect-mediated diseases began to occur in earnest only after many populations began to establish permanent settlements since the beginning of civilization.

Parasitic diseases were also among the diseases that threatened the health of the people as they settled down. To survive, parasites must be able to grow from larvae to adult forms in their host or mediator; as such, they are able to propagate by taking human bodies as hosts and infecting them constantly. To meet these preconditions, however, a large number of populations should live in one area for an extended time. Therefore, it was difficult for parasites to continue to act as infectious agents for the hunter-gatherers, who had small populations of about 30–50 people, each moving frequently. Thus, infestation could occur during the period of hunter-gatherers only if the parasites inhabiting the soil or animals accidentally infected humans or if the period of inhabitation in the human body accounted for a significant proportion of the parasites' entire lifespan.[4]

Some parasites began to use domestic animals as the intermediate hosts after the introduction of farming and herding and come in contact with people later, causing diseases. Parasites such as tapeworms take cows and pigs as intermediate hosts, but they can also be transmitted to humans and cause parasitic diseases. Once people eat the meat from cows or pigs infected with tapeworms, they are not only infected with the parasite but the eggs of the tapeworms are excreted from the human feces, setting the circulation loop in motion, in which the cows or pigs again eat foods containing the said human feces. Perhaps during the age of the hunter-gatherers, before cattle began to live close to people (i.e., before the start of herding), the parasites that infected animals and humans alternately would have existed infrequently. Ever since humankind entered the agricultural era, however, parasitic diseases have spread far and wide as parasites frequently take the domestic animals as their intermediate hosts. Therefore, it can be said that parasitic diseases increased in earnest only after civilization began and people began to settle down with domestic animals.

As farming enabled the harvest of increasing amounts of surplus agricultural products, the farming products were stored, which became a condition for animals like rats to flourish. Rats, in particular, are among the animals that can adapt most quickly to the changes in the people's residences. They can eat the food waste and grains stored in people's houses without difficulty, and the people's houses protect them from their predators, making it easy for them to survive and reproduce. Thus, in the early civilizations, the rats flourished in almost every place where people lived. On the other hand, animals such as wild birds and wild mice, which usually do not live near human habitats, could also get food increasingly easily near the people's houses, thereby increasing their indirect contact with people. As the wild animals began to appear increasingly closer to people's habitats during the time of the early civilizations, they could spread pathogens to humans via the animals that were in direct contact with the humans, such as the domestic animals or rats. Thus, the diseases caused by the transfer of pathogens from animals to humans also increased in earnest only after civilization began.

Livestock breeding changes the disease pattern of humankind

During the past 50,000 years, humankind was no longer confined to Africa but began to spread and settle all over the world. This period had a significant influence on the genetic characteristics and environmental adaptability of humans. It can be said in particular that the characteristics of the disease or immunity status as well as the biological characteristics of each race were decided during the period leading up to the beginning of civilization, when humans started both farming and herding. As the cultivation and irrigation facilities for agriculture were being made, the surrounding living environment was greatly changed. With the introduction of livestock, animals were brought into intimate contact with humans, prompting people to establish a new kind of relationship with germs from animals. In other words, there was a challenge for humans to adapt to their new environment, and only those who succeeded in such adaptation survived in their respective settlements.

As the climate became warmer in the mid-latitudes, the first civilization grew as farming began in the Fertile Crescent located in the east of the Mediterranean Sea. Then, civilization spread mainly to the regions in similar latitudes where the climatic conditions, such as the temperature, were relatively within a comparable range, thereby prompting the development of civilizations along the east–west axis: Mesopotamia, Egypt, India, and China.

Farming-based civilization was accompanied by the domestication of animals, and in most cases, domestic animals were used as a means of supporting farming while herding itself, in some occasions, replaced farming completely. As such, the raising of domestic animals marked the beginning of civilization with agriculture, but it also meant that human diseases began to develop to a new level. In the age of the hunter-gatherers, animal hunting did not end up with having close contact with animals, but raising domestic animals after the advent of civilization meant that the animals had more chances to transfer germs to humans.

Ultimately, animal domestication provided a hotbed for various infectious diseases. This is because even when humans are not the normal hosts to pathogenic germs, their intimate contact with animals increase the likelihood that the mutant germs in the animals may take humans as hosts. In addition, the germs that took humans as hosts were able to spread easily among people due to the people's clustered habitation lifestyles. Ultimately, these conditions combined to cause new diseases in humans. Goats brought the *Brucella* disease, and contact with cows caused the highly lethal anthrax as well as smallpox and diphtheria, whereas pigs and chickens brought influenza, and horses brought colds.[5] The vast majority of other viral diseases also appeared after the introduction of animal domestication.

Meanwhile, the geographical conditions of the Americas spreading along the north and south axes, with significantly different climatic environments, made it difficult for humans to smoothly propagate farming and herding owing to the limitations imposed by the natural conditions.[6] In other words, the method of domesticating a wide variety of animals in Europe and Asia did not spread easily in the Americas, where the climate varies significantly. Instead, only a few animals, such as llama or alpaca, were domesticated in the Americas. These animals were not used for farming but mainly for carrying goods, in addition, to obtaining their fur.

As those who have suffered from some disease, however, became immune to the epidemic, the populations that had been infected with various infectious diseases differed greatly in their collective immunity characteristics compared to those who had never been infected with the disease. The inhabitants of the Americas had less experience of being infected with the communicable diseases that had infected their European and Asian counterparts due to the former's limited animal domestication, and thus, their ability to collectively immunize themselves against such communicable diseases was inferior. This was the root cause of the epidemic outbreak of infectious diseases among the American natives in the early days of the Europeans' conquest of the Americas.

This phenomenon became even more dramatic in isolated areas, such as Polynesia. Measles is a disease that was caused by a virus that had taken a cow as its host, but the virus became mutated and then took a person as a new host. No Polynesian had been infected by the measles virus until the Europeans visited Polynesia in the Age of Exploration, as the Polynesians had migrated to the various islands scattered around Polynesia before cattle were domesticated thousands of years ago. Eventually, the measles virus transmitted by the Europeans when they conquered Polynesia caused massive deaths among the Polynesians, who had no immunity to measles.

Is disease a punishment from god?

Although there are descriptions in the Babylonian or Assyrian literature of the early diseases that they suffered from or the medicines that they took for such diseases, the Egyptians kept much more detailed records about them. Therefore, to some extent, we can get a glimpse of the illnesses and medicines of ancient Egypt. According to the records, there were physicians as well as medical treatments for diseases in ancient Egypt at around 3000 BC. The Egyptians, however, thought that illness was caused by bad souls or poison entering the body, and as such, they either prayed to the god of healing or treated the illness by taking medicines prepared with herbal ingredients. As such treatments were usually spell, magic, or ceremonial rituals, the clergy often served as physicians. For instance, Imhotep was a priest of Pharaoh, but he also acted as a physician and was later worshipped as a god of medicine. In ancient Egypt, there were physicians who specialized in the eyes, teeth, or abdomen, suggesting that various forms of illness had been recognized and treated at that time.[7]

The disease that afflicted the ancient Egyptians most seriously was schistosomiasis, a disease that causes the symptoms of abdominal pain, diarrhea, and hematuria by infecting intestines or the urinary tract after a freshwater parasite, schistosome, enters the human body when people eat snails, which are host of the parasite. As the Nile River overflowed for several months every year in Egypt, an environment conducive to the spread of snails was created, prompting many people to consume snails as food. Therefore, schistosomiasis infection through the consumption of snails became prevalent, resulting in anemia, infertility, or even deaths. In addition, malaria and other diseases caused by mosquitoes broke out periodically. Infectious diseases like smallpox, dysentery, and typhoid also broke out frequently in the Nile River basin, which was densely populated.

In ancient China at around 2000 BC, when the Shang Dynasty, which represented the Bronze Age, was established, many things began to be recorded, making it possible to grasp their lifestyle and culture. Among these, the records of diseases engraved on the back skin of turtles show that they observed the symptoms, causes, and progress of disease in detail. They diagnosed and classified various diseases, such as skin diseases, boils, stomach pain, and pediatric, obstetric, and mental diseases. The engravings on the back skin of turtles not only describe insect-mediated diseases such as malaria and scabies but also show the time when specific diseases broke out, or the epidemic patterns of diseases. In particular, the Chinese letter <°>, which means "heart," has a real heart shape, suggesting that the Chinese had some anatomical knowledge as early as at that time.[8]

What is notable in Chinese medicine is the use of acupuncture. Acupuncture was used in China from the era of hunter-gatherers, well before the beginning of agriculture. At first, they performed a simple procedure of making an incision on a boil using a sharp stone tool. Later, they came up with a thin needle made of bone or bamboo. As acupuncture developed to a considerable level, it was often observed that the body area to which a needle was applied and the area where the symptoms improved were different, leading to the idea that parts of the human body are interconnected with one another through the flow of energy (Qui). This idea soon evolved into the medical thinking that there are "meridian systems" connecting the entire body organically and that the flow of Qui keeps the entire body healthy through such meridian systems. This idea, however, was not developed systematically along with scientific development; instead, what had dominated the public perception of disease at that time was shamanism. The vast majority of people believed that an angered ancestor cursed his descendants, making them sick or inviting an evil spirit to enter their bodies to make them ill. Therefore, the people believed that the illness would be cured only if they appease their ancestors or ask them to expel the evil spirit in a shamanistic ritual. With the exception of the partial use of acupuncture, the treatment methods for diseases were mostly ceremonial as such.

As the ancient civilizations of the Americas did not leave enough records, unlike the ancient Egyptian or Chinese civilizations, it is difficult to know much about the diseases that inflicted them and their medical skills before the Europeans entered America. As the Americans, however, shifted from the lifestyle of the hunter-gathers to an agricultural lifestyle as we observed in other parts of the world, they also changed their dietary habits from eating a variety of foods from hunting and gathering to a limited number of foods, such as potatoes, thus disrupting the balance of their nutritional intake compared to the era of the hunter-gatherers. Archaeological excavations revealed that there was more iron deficiency during the transition period from the hunter-gatherers' lifestyle to the agricultural lifestyle, suggesting that the intake of various nutrients was not sufficient during that period.[9]

It can be concluded based on the lifestyle of the time of their ancient civilizations that the major health threats to the people living in the Americas during the time were malnutrition, parasitic infections, physical injuries, and poisons.[10] On the other hand, the remains unearthed in areas such as Peru show that the drilling of human skulls had been practiced extensively. It may have been a procedure to treat diseases like headache or epilepsy or a practice with the magical purpose of exorcising evil spirits from a man possessed by them. Perhaps the exorcism of evil spirits and the treatment of a disease were the same, and the person who played the role of a therapist was also a physician-cum-shaman.[11]

1.2 Humankind Begins to See Disease Through the Eyes of Reason

Along with the philosophical reflection on human existence, humanity began to view sickness and death with rational eyes, veering away from the religious belief that considered them punishments by the gods. Hippocrates was at the forefront in pioneering such new thoughts, followed by Galenus. Professional physicians like Bian Que in China and Sushruta in India also appeared, and the profession of physician started to form. As such, Europe and Asia have made considerable medical progress compared to the early civilizations. Despite such developments, however, there was no way to stop the arrival of the age of epidemic.

Reason leaps forward

Karl Jaspers, having observed that humanity's leap of reason took place simultaneously across various areas, such as Greece, India, and China, from the eighth century BC. to the second millennium BC, named such era "Axial Age." During this period, science and philosophy in Greece, political philosophy in China, and philosophical thought in India flourished, each established as the central idea of its culture.[12] Therefore, this period was marked by the blossoming of human reason, which was required to lead the already-complicated social system then, along with the rise of reflective thinking about what human existence means and what is the ideal society. Thus, it can be said that such a leap of reason took place after the early agricultural society successfully evolved into a complex social, political, and economic system of the ancient cities.

During this period, the foundation of modern civilization was established along with the general development of human culture, including literature, politics, art, and religion, all based on the questions and philosophical teachings on the meaning and purpose of life, suffering and death, and good and evil. In fact, the emergence of these great ideas also suggests that the framework of a society that is capable of accepting enlightenment and great teaching was laid out widely, at least in the European and Asian cultures. However, great advances did not spread from the central culture to other cultures but appeared relatively independently in each region in almost the same period, to our astonishment. In this period, cultural exchanges with the people in remote areas were not supposed to be active even though contact and trade between population groups in different areas were not blocked completely. It is therefore reasonable to assume that the foundation conducive to the creation of such ideas was not propagated from one place to other regions but was already in place across the wider regions of the Old World, particularly in Europe and Asia.

Ever since the birth of the ancient city-states, the development of the metal and letter systems has greatly improved people's consciousness and cultural level as well as human productivity. As people's productivity and level of consciousness improved, demands for new social structures and systems emerged, prompting the need to find a new social order by breaking down the preexisting hierarchy that had led to the creation of civilization. There was a swirl of political changes and wars behind the birth of this great philosophical thought, which was prompted by such demands. In the meantime, people came to think more deeply about themselves and the world that they had been accepting without any question in the past, arriving eventually at a philosophical contemplation of universal people or the humanity, beyond individuals and groups.

In particular, ancient Greece was the region where a political system, democracy, had been created based on the reflection of the relationship between humanity and the world, and where a great leap forward in scientific knowledge, such as in mathematics, astronomy, and philosophy, had transpired. Not only did ancient Greece make great advancements over the past achievements of the ancient city-states; it also laid the groundwork for today's politics and science. In fact, science began with an attempt to accurately count the number of objects and measure the sizes of arable lands, and to predict the weather and climate by observing the movements of the sun, moon, and stars. In particular, astronomy and mathematics made a significant leap in ancient Greece, although they began to develop in ancient Egypt and Sumer. The efforts to collect the public opinions and to pursue publicly shared knowledge in ancient Greece not only increased the overall amount of knowledge but also prompted the development of scientific disciplines that aspired to verify such volume of knowledge. With the development of astronomy and mathematics, the desire to discover the ultimate principle of natural phenomena that had been taken for granted led to the development of philosophy, which began as the study of the answers to the question "What is the origin of all phenomena?"

The political and social turmoil that swept through ancient Greece during the Persian War was a great ordeal for the city-states in the region, but it was also a crucial opportunity of ancient Greece for a leap forward as the war brought its potential to the forefront. In particular, Athens, which deployed its naval forces to defend the city-state, emerged as the center of the flowering culture prompted by the political demands for democratization by the people who participated in the war, coupled with the wealth obtained from the alliance that had been built. The citizens of ancient Athens enjoyed a decade of peace after the Persian War, but their existing values and order were destroyed again after Athens lost in the Peloponnesian War, which lasted for 30 years. In such situations, the new philosophical thought was created centered on Athens, thereby gradually eliminating the religious beliefs and traditional values that depended on gods, and creating the demand for new values and a new social order that would replace them.

Greek philosophers such as Socrates, Plato, and Aristotle emerged in the period of such chaos and change, and they laid a rational basis for philosophy by reflecting earnestly on human existence and interpreting the human relations with the world. Interestingly, massive wars unprecedented in scale in China and India also raised questions about life, death, and the life after death, laying the foundation for the rise of a great philosophy or religion. In China, great thinkers like Confucius, Mencius, and Laozi emerged and taught on ethics and morality, while in India, the writers of the scripture Upanishads preached about life, death, and ultimate truth.

In addition to the philosophical reflection on human existence, humanity began to view sickness and death with rational eyes, veering away from the religious belief that considered them punishments by the gods. Hippocrates was at the forefront in pioneering such new thoughts, followed by Galenus. In China, professional physicians like Bian Que appeared, and in India, a distinguished physician named Sushruta emerged, and with these changes, the field of medicine was put in order and the profession of physician started to form.

Hippocrates brings down disease from god's domain to the human domain

Hippocrates, who was born in Greece in 460 BC, was one of the most important figures in the history of medicine. While Plato and Aristotle laid the foundation for philosophy through their deep reflections on humankind and the world, Hippocrates, who lived contemporaneously with them, laid the foundation for medicine by deepening understanding of the causes of disease through observations of humans and diseases. Born in the Greek island of Kos, Hippocrates studied medicine at Asclepeion in Kos Island. Asclepeion was the temple of the god of medicine, Asclepius, where the patients went to have their diseases treated. Hippocrates is said to have lived close to 90 in the temple, teaching medicine to his disciples and practicing it to his patients after learning about illnesses and cures.

In his article "About Air, Water, and Places," Hippocrates explained how health conditions and disease patterns may differ according to the living environment. In the first part of this article, Hippocrates explained that a person's living environment determines his health condition by influencing his characteristics and humors. In the latter part, Hippocrates describes how the racial characteristics, such as those of the Europeans and Asians, may differ according to the given environment.[13]

For example, he explained that diseases such as sinus cold, dysentery, acute fever, and paralysis are closely related with the seasonal changes, while the personality and body shape are determined by the geographical characteristics. He often explained the cause of illness and its countermeasures in details. In his description of urinary stones, for instance, he suggests that the sticky and turbid part of the urine condenses and gets crystalized before growing progressively larger and becoming urinary stones. Therefore, he recommended that patients drink plenty of diluted wine to treat the disease, which is virtually no different from the recommendations of the contemporary physicians: "Drink enough water" to prevent and remove urinary stones.

The fact that Hippocrates considered air, water, and places as factors that affect the health is related to the development of Greek philosophical concepts. Thales, who is largely considered the creator of Greek philosophy, claimed that everything in the world is made up of water after discovering shellfish fossils on the rocks on the hills. On the other hand, his follower Anaximenes of Miletus thought that the air in the sky had descended to become water, and that the earth and stones were created by pressure. The worldview of Thales and his disciples is that all things in the world are made up of water or air.[14] As humans are among the things existing in the world, it was a natural consequence for Hippocrates to inherit this philosophical concept and to conclude that the living environment, such as the air, water, and land, which compose the world, have a critical influence on human body and health.

Therefore, it can be said that the development of medicine was influenced by the philosophy of the time, although medicine also had a considerable influence on philosophical development. Indeed, medicine and philosophy as they existed at that time cannot be understood as separate academic disciplines, independent from each other, as they are today. Plato, for example, referred to Hippocrates many times in his *Dialogue*, which states that those aspiring to build a nation should consider the environmental factors that can affect the flesh and minds of the residents. As Plato and Hippocrates were contemporaries, they not only developed philosophy and medicine, respectively, but also had a considerable influence on each other.[13]

At that time, the thoughts and ideas derived from ancient civilizations like Egypt and Mesopotamia, which assumed that disease results from the anger of or punishment from the gods, were still prevalent. Separating diseases from the territories dominated by the gods, Hippocrates argued that diseases are caused by environmental factors and dietary and lifestyle habits. In other words, he brought disease from the divine domain to the human domain. In addition, he adopted a nursing-oriented clinical approach to make his patients better by improving these factors and allowing the patients to rest, most of which resulted in great success.

Not only did he closely monitor the patients' pulses, fever, pains, and movements; he also paid attention to the patients' environments and families. Such approach to disease or the patients is no different from today's approach. Due to his lack of knowledge of anatomy or physiology, however, Hippocrates could not go beyond the limits of explaining human temperament and disease based on the action of the four humors: blood, phlegm, yellow bile, and black bile. Such an idea on illness is not scientific or practical from the view of modern medicine, but the idea that illness occurs when the balance among the four aforementioned humors breaks down is significant in that it laid the foundation for understanding that disease is not merely a result from relationship between cause and effect but can also be seen from the perspective of harmony.[15]

In fact, the theory of the four humors originated from Empedocles of ancient Greece, and Hippocrates adopted it later, establishing it as a major theory of pathological development, which prevailed for a long time. According to the theory of the four humors, humans have four different humors, each with a different effect on human health and behavior: (1) blood, which is warm and invigorating; (2) yellow bile, which is dry and brave; (3) black bile, which is cold and depressing; and (4) mucus, which is moist and slows down a person's behavior. The theory states that everyone has a mixture of these four humors, and that the health of each person depends on the extent to which the balance among these humors is maintained in the person's body. In other words, the theory claims that people become sick when the balance among these humors is broken. Therefore, restoring the balance in one's body is an important treatment for illness. Especially, the main therapy of Hippocrates involved waiting and providing good food to the patient, based on the assumption that people are born with a healing ability by restoring the balance that can help them recover naturally. However, as the ancient Greeks did not accept the idea of dissecting the human body, Hippocratic medicine was only to be limited to providing care and judging the prognosis based on the observation of a person's superficial conditions.

Rising to the center of western medicine through anatomical knowledge

Born in Greece in AD 129, Aelius Galenus studied ancient Greek philosophy, including the Stoic and Epicurean philosophies, before studying medicine in Asclepeion, the medical education institution at the time. He then went back to his hometown Pergamon as a physician and took care of the "gladiators," who were trained to fight with weapons against other men or wild animals in an arena. Having observed the gladiators' injuries and recovery process, he understood the importance of food, physical fitness, and hygiene and learned the human anatomy and the treatments for fractures and injuries. He further studied medicine and performed medical experiments for more than 9 years in Alexandria, Egypt. In particular, the human skeletons at Museion, a museum of Alexandria, provided him a valuable opportunity to gain knowledge about the human skeletal structure at the time when the dissection of the human body was prohibited, and the dissection of pigs', oxen's, and monkeys' bodies allowed him to complete his theory on the human body.[16]

Although Galenus acquired his knowledge of the human anatomy by dissecting animals, he arranged such knowledge relatively well, with the exception of some errors, and it later became the widely accepted anatomical knowledge of the human body for a long time. His critical error was that he assumed that the blood from the liver migrates to the right ventricle of heart and then partly to the lungs, with the remainder flowing to the left ventricle. Although the ventricular wall of heart is so thick that it is not possible for the blood to move through it, Galenus explained that the blood travels through tiny holes in the ventricular wall. He also believed that the role of the lungs is to cool the hot blood by providing the heart with air and to blow the energy of life ("pneuma") into the human body. This blood circulation process, which was summarized as such by Galenus, had in fact been raised earlier, in the third century BC, when Praxagoras of the Alexandrian School observed that the carcass's artery was empty while its vein was full of blood. Galenus also assumed that the artery supplies the air that is sucked into the lungs, then to the whole body, and that the veins serve as conduits for supplying nutrients to the whole body after food changes into blood in the liver.[14]

Galenus went to Rome later and there worked as the attending physician of Emperor Marcus Aurelius and his successor, Emperor Commodus. Whenever he had questions about a disease, he accumulated knowledge by dissecting animals, and he developed various treatments, such as suturing wounds and ligating the blood vessels. His theory may not be acceptable from the perspective of modern medicine as he divided people's temperament into four groups based on the theory of humors due to the influence of Hippocrates. Still, however, it should be acknowledged that he had substantial accomplishments, including the establishment of the bases for anatomy and physiology by dissecting monkeys' and pigs' bodies and experimenting with the nerve blockade. Galenus' idea then became the central theory of Western medicine for more than 1000 years, and Galenus' authority came from his experience of animal experiment and from observing the inside of the human body while treating gladiators. The claims of Galenus, which were based on the limited anatomical knowledge existent at the time when dissecting the human body was strictly forbidden, became absolute beliefs even though their scientific veracity was somewhat dubious, because they could not be proved false at that time.

Galenus also wrote quite a number of books, which allow us to understand his thoughts on health and disease. In his book *On the Theory of Hippocrates and Plato*, he argued that there are three kinds of souls in each organ: (1) rationality in the brain, (2) spirituality in the heart, and (3) appetite in the liver. He established the idea that a person can be healthy only when each of his organs and his soul function well independently. The notions that the body and the mind are indistinguishable from each other and that one can be healthy only when both his body and mind function well were revolutionary ideas at that time. Galenus also described the principles and methods of psychotherapy that reveal and treat the deep inside of the patient in his book *On the Diagnosis and Treatment of Mental Illness*. As such, Galenus perceived mental disorder as important disease as well as physical illness. He understood that mental illness is not caused by evil spirits or God's curse but by passion or the secrets hidden inside the patient.

At the time when Galenus was active, however, the idea that the course of disease is determined by divine power or revelation had still dominant influence among the physicians. Among these physicians, Galenus argued that medicine can be understood as an academic discipline that should be carried out by combining theory, observation, and experimentation, and that accurate diagnosis can be used to determine the course and prognosis of a disease, thereby contributing greatly to the development of medicine as a legitimate science. It can be said that Galenus succeeded in formulating scientific ideas by fusing the philosophical thoughts that had a great influence on the society at the time, such as Plato's rationalism and Aristotle's empiricism, within the scientific framework of medicine.

Galenus could have been successful as a physician partly because of the illnesses prevalent during his time. The main diseases at that time were injuries and fractures caused by wounds and fights, infectious diseases like malaria that are mediated by insects but do not spread directly among people, contagious diseases like leprosy or tuberculosis that do not spread extensively even if they spread among people, or jaundice or epilepsy that often occurs in isolation, and the vast majority of such diseases can be cured to some extent, with surgical treatments, stabilization, or dietary management.

There has been a significant change, however, in the disease pattern. Epidemics began in the ancient Roman Empire, under the rule of Marcus Aurelius in AD 164. The epidemic that began to spread among the Roman troops during the war was a terrible disease that even Galenus could not understand and manage. The Roman Empire had a vast territory, but on the other hand, it was under constant attack from outside, and to defend the Empire, the Roman troops were stationed to the east of Iraq today and to the west of the Rhine. Then, a plague struck the Roman troops who had been dispatched to Syria to quell the rebels, leaving many dead, and by AD 166, the plague had reached Rome.[17]

Galenus himself experienced this epidemic in a military garrison, where he was able to observe the rashes, blisters, pustules, and bloody diarrhea, as well as high fever. According to the records left by Galenus, we can assume that it was smallpox. This epidemic led to the death of 2000 people a day in Rome at its peak. The power of this epidemic had a considerable impact on the ancient Roman economy as well as on the military conscripts. Moreover, as there was no proper way of dealing with the epidemic, the past perception of diseases that were punishments by evil spirits or gods resurfaced and spread again. In the end, Rome was hit by infectious disease epidemics as well as by military attacks from outside, which triggered the decline of the Roman Empire. The medical theory and practice also stopped developing in its wake, before Europe stumbled into the long Dark Age.

Oriental medicine took a different path from western medicine

Whereas the theory of four humors proposed by Hippocrates and Galenus had been the basic theory of pathological development in the West until the dawn of modern medicine in the late 18th century, the yin-yang and five-elements theory, which sees the human body as a semblance of the universe, had been the basic theory of pathological development in East Asia until the introduction of modern medicine from the West. According to the yin-yang and five-elements theory of disease, disease is caused not by bad spirits or ghosts but by the breakdown of the order of the human body, which is made up of yin and yang and five elements (i.e., tree, fire, soil, metal, and water). In other words, disease is seen as occurring when the balance between the five organs belonging to yin (i.e., the heart, liver, lungs, kidney, and spleen) and the five other organs belonging to yang (i.e., the large intestine, small intestine, gallbladder, stomach, and bladder) is broken.

Unlike Galenus, who developed the mechanistic notion, although based on the theory of four humors, that disease is caused by an abnormality in a specific body part and can be cured when the abnormality is corrected, Bian Que, who is largely considered representing the East Asian medicine, viewed disease from the perspective of an organismic concept based on the yin-yang and five-elements theory, which assumes that the body parts are integrated to reflect the entire body. The mechanistic theory of Galenus and the organismic theory of Bian Que are in line with the central framework of the thoughts shared by Western and Oriental medicine today, respectively. In other words, it is understood, according to the yin-yang theory of disease, that a disease occurs because the meridians connecting the body organically are blocked, hindering the smooth flow of energy via the meridian. Given that the meridian cannot be observed anatomically, this idea cannot but be considered baseless from the perspective of Western medicine, which understands disease with medical logics based on observations and experiences gained from experiments.

The book *The Inner Canon of the Yellow Emperor* contains medical thoughts formed from the Spring and Autumn/Warring States (SAWS) Period to the Qin Dynasty in China, which are known to have been compiled by the book's unknown author from the conversations that the emperor had with his physicians about illness and medicine. In particular, a distinguished physician called "Qibo," when asked by the emperor why people were not healthy, said, "People get sick as their lifestyle and habits change," implying that people can stay healthy as long as they control their food intake, live a regular life, do not use force beyond their capability, and keep their body and mind in harmony with each other. To summarize the basic ideas of the said book, the yin and yang are the fundamentals of life, and one would lose his/her health when the harmony between yin and yang is broken. In other words, the theory proposed by the book is that the relationship between yin and yang is maintained at delicate balance even though the relationship between the two may change over time and that the principle of yin-yang is contained in the body structure, physiological functions, and pathological changes in the human body; therefore, it can be applied to the diagnosis and treatment of diseases.[18] The book's author argues that people can live a healthy and long life if they maintain the inner harmony of their body as well as harmony with nature.

Just as the Persian War and the Peloponnesian War triggered a philosophical leap forward in ancient Greece, the rapid social changes and cruel wars in the SAWS Period in China, combined with the advancement of science and technology, laid the foundation for the ideological flourishing called "100 Schools of Thought." Until the period when the Zhou Dynasty lost its power, pushing China into the chaos of the SAWS Period, China remained relatively stable, supported by the gradual growth of its productivity. At that time, the ancient Chinese understood disease as almost every other ancient civilization did, as being caused by the gods' wrath or by the curse of evil spirits. The chaos of society that prevailed during the SAWS Period, however, brought about the leap of human reason in China, and the concept of health and disease also developed by the theory of yin and yang. As Emperor Qin unified China, ending the SAWS Period and beginning the Qin Dynasty, the empire began to impose strict control to ideas by pushing for centralized governmental oversight through the burning of books and burying of scholars. The liberal culture, therefore, that flourished in the SAWS Period could no longer bloom at this time, and Chinese medicine also entered a long period of stagnation, failing to evolve beyond the theory of yin-yang and the five elements.

India was divided into several countries after the end of the reign of King Asoka, who ruled India in the third century BC. As the trade with the Roman Empire increased and the culture of each country developed, India entered the golden age of cultural bloom. From the beginning of the Gupta Empire in the fourth century until the fall of the empire in the sixth century, there was a great leap forward in all aspects of civilization and across all academic disciplines, including philosophy, literature, science, technology, and religion. The physicians of India, who had been under the direct influence of the Roman Empire's medicine, no longer regarded disease as the result of an evil act but understood it to be caused by the imbalance of the humors. For instance, Charaka, one of the physicians representing Ayurvedic medicine in India, argued that the best treatment for any disease is its prevention and that one's health can be maintained if the balance of the three humors (i.e., bile, sputum, and breath) is likewise maintained.[19]

Ayurveda, which means "knowledge of life," is the representative medical knowledge system in India and has been developed by adding new theories and therapies to the existing knowledge on herbs that had been handed down traditionally. The detailed examination of patients, the diagnosis of diseases, their treatment, and prognosis were also laid out in the books of Ayurvedic medicine, which is more practical for clinical application than Galenus or Chinese medicine at that time. For instance, Schusuruta, a physician, introduced various advanced surgical procedures, such as cosmetic surgery of the nose and ear and cataract surgery, and identified and classified many different diseases. In addition, Ayurvedic medicine taught students not only anatomy but also physiology and pathology, by far the most advanced medical knowledge ever anywhere in the world at that time.[20] As such, the civilization centers of Europe and Asia, including India, have made considerable medical progress compared to the early civilizations or ancient city-states, although they became stagnant afterward. Despite such developments, however, there was no way to stop the arrival of the age of epidemic.

1.3 Epidemics Changed History

Conditions conducive for an epidemic to spread were well established, along with development of civilization. The power of the epidemic continued to grow, and it became a subject of horror as people died hopelessly. Once again, disease and death became divine destiny that reduced humans into entities hopelessly awaiting god's disposition. In particular, smallpox was a terrible disease with a high mortality rate when infected. However, even the power of smallpox was dwarfed by the massive devastation inflicted by the Black Death. As Europe underwent an epidemic of the Black Death in the 14th century, the feudal system was dismantled, and the religious beliefs that sustained the society were shattered, highlighting the need for new thinking and a new order. In contrast, Asia did not experience such a terrible epidemic, which was one of the reasons for lagging behind Europe in terms of development afterward.

Conditions ripe for epidemics of infectious disease

As civilizations developed, city-states evolved into empires based on a string of cities that started to form along the trade routes, which provided very good conditions for the propagation of pathogens as well as for the exchange of goods.[21] With the increase of the commerce and trade volumes, cities gradually grew enormous, and their political systems became imperialized. Furthermore, the roads and wagon wheels were standardized to facilitate the rule of the empires, and the increasingly convenient transportation network helped infectious diseases spread easily. In particular, the Roman Empire built roads crisscrossing its territory to improve the convenience of travel, which on one hand made it easy for infectious diseases to spread.

Populations became much more concentrated than ever before while livestock and animals came into closer contact with people. However, sanitary facilities capable of treating the manure of humans and livestock were not in place, setting the conditions conducive for microbes to flourish. In the end, not only did it become possible for a plague to occur within the residential areas; locally confined epidemics previously also became fully prepared to spread along the trade routes. Moreover, the cities and empires constantly waged war against one another as they acquired wealth through conquest as well as commerce and trade, expanded their respective territories to accumulate wealth more, and needed slaves to replace their labor, thereby triggering explosive growth in military activity, trades, and slave movements. As a result, conditions conducive for an epidemic in one area to spread to other areas were well established.

For infectious diseases to become prevalent, there must be livestock breeding, densely populated residential places, and active trade and exchanges. On the other hand, however, the microorganisms causing infectious diseases must be strongly virulent and the resistance of their new hosts to the germs must be weak. As civilization had not developed on a large scale before microorganisms and humans created new epidemics, microbes had not had a chance to adapt to their new hosts via natural selection; nor had humans been given an opportunity to equip themselves with an immune system strongly enough to fight infectious agents. Therefore, microorganisms could spread from one person to another by attacking their hosts with powerful virulence, and there was no way for the humans then to stop this spread.

In particular, the conquest of a new land or development that changed the natural environment continued to provide those microorganisms that had not yet taken humans as natural hosts an opportunity to do so. The microbes that had successfully mutated and had taken humans as a new host brought in a new epidemic among the people. Such change of hosts was likely to have also occurred during the hunter-gatherer and the early civilization periods, but it was usually very limited at the time and could not have caused a massive epidemic. In other words, even if microorganisms caused disease at that time, they failed to spread as an epidemic and returned to their original hosts. However, as the conditions conducive for spreading an epidemic were in place in the era of city-states and empires, microorganisms could cause an epidemic of infectious diseases by taking humans as their new hosts.

When the pathogenic bacteria enter a human body for the first time, they could kill the infected persons with a strong virulence, but sometimes the infected persons gain immunity and recover from the disease. When the number of people who had recovered from the disease through immunity increases in the population, it becomes harder for the epidemic to spread among such people, just as it gets more difficult to break through the barriers when there are an increasing number of barriers. Therefore, if a new population that has no immunity does not keep flowing in or if babies are not born in sufficient numbers continuously, the infectious disease epidemic will die out in the end.

On the other hand, in areas where the aforementioned conditions ripe for the spread of epidemics are being created constantly, infectious disease epidemics can become prevalent for an extended period of time. Cities with large populations can meet such conditions as there are many children therein. Children are more likely to be exposed to unsafe environments because they are often concentrated in schools and daycare centers, staying together in a room with poor ventilation, particularly in the winter season. In addition, an infectious disease can spread among them easily because the children are not yet equipped with a strong immune system. Another reason that large cities are well suited for infectious disease epidemics is that the pathogenic germ can continue to infect people if the population is very large, because there is a large pool of people with insufficient immunity so that the infection does not easily go away. On the other hand, the epidemic would sweep over the small communities, resulting in either the death of most residents or their eventual attainment of immunity. Therefore, if there are no more people to infect in the community, the bacteria that caused the epidemic in the first place would die out as well or survive only in the case of transmission to other community.

The power of the epidemic continued to grow along with urbanization, and once an epidemic began to spread, it became a subject of horror as people died hopelessly. There had been attempts to find the cause of the disease in the living environment, such as in the air, water, or animal carcasses, in an effort to prevent its root causes. Such efforts, however, were not very successful, and the fear of death made the people dependent on the gods again. The Christianity and Islam, which had been active throughout the Roman Empire and the Persian regions, were expanding their power based on people's fear of death. Once again, disease and death became divine destiny that reduced humans into entities hopelessly awaiting god's disposition.

Smallpox brought the fear of epidemics

Smallpox is a disease caused by infection with a virus called "variola." It usually spread in densely populated places because the variola virus can spread to other people through the small droplets discharged from the upper airway when an infected person cough or sneeze, as well as through the pus from the skin pustules, or the scabs falling off the skin. When a smallpox virus enters the body, it could cause only a mild fever if the host had strong immunity. The majority of viral infection, however, often spread to the entire body, causing serious illness showing various symptoms, like hemorrhage and skin pustules. To conclude, smallpox was a terrible disease with a 30%–50% mortality rate when infected.

The virus is believed to have taken a monkey or a squirrel as a natural host before migrating to a human body and causing smallpox. Perhaps the smallpox viruses in the African jungles jumped to humans, mediated by monkeys, before entering the Europe and Asia along the trade routes by piggybacking on the slaves, merchants, soldiers, or explorers. The smallpox virus was also found in ancient Egyptian mummies, and Ramesses V was reported to have died from smallpox in 1157 BC.[22] Ever since the arrival of smallpox from ancient Egypt, India had been struggling with its chronic outbreaks for over 2000 years, and it arrived in China in around 250 BC, when the Huns living in Central Asia attacked China.[23]

Smallpox first broke out in the Roman Empire in AD 164. and then spread to Arabia on a massive scale during the Elephant War, when Ethiopian troops attacked Mecca in AD 569. It then migrated to Western Europe from the Middle East through the crusaders and pilgrim, including Italy, France, and Spain, turning Western Europe into an endemic area of smallpox. Over time, it spread to northern Denmark, England, and Greenland, and then to as far as Russia by the 17th century. The most important reason for such massive spread of smallpox was that these regions saw growing urbanization, with more people living in a more concentrated way than ever before, promoting commerce and trade as well as war.

Once Europe became the endemic location of smallpox, the colonization efforts by the Europeans in the Era of Exploration from the 15th century played a role in spreading smallpox to the New World, which had so far lived independently from the Old World. In 1507, the smallpox virus was spread to the Caribbean islands by the Spaniards, thereby causing the first outbreak of the disease in the New World before spreading to other parts of the Americas. In particular, when the slaves were sent from West Africa to Cuba to work in the mines, the smallpox viruses came in by piggybacking on the slaves before spreading rapidly across the American continent. When the Spanish army led by Cortes arrived in America in 1519, smallpox had already swept through the Aztec Empire even before the start of the war, killing nearly half of the native population. This enabled the Spanish soldiers, who numbered just a few hundreds, to easily conquer the Aztec Empire in 1521. Smallpox also swept the Incas, and when Pizarro of Spain arrived in 1532, it was already in chaos, facilitating the easy entry of the Spaniards into the Inca Empire.

The reason that the Spanish army was able to defeat the Aztecs and Incas with ease was that the natives of these empires had yet to attain immunity to the smallpox virus by the time the Spanish army arrived because they had never been exposed to the virus until the Europeans brought it in from the Old World. About a century later, the smallpox viruses that the Europeans brought to South America began to spread to North America. Some Europeans even spread smallpox on purpose by giving smallpox patients' blankets as gifts to the North American natives who refused colonization, fully knowing that smallpox could be transferred through the blankets used by the smallpox patients. Smallpox then spread to Australia in the 18th century, and to the rest of the world in the 19th century, via the baggage, mails, crews, and passengers aboard ships.[24]

The black death ends feudalism and makes powerful nations appear

Even the power of smallpox was dwarfed by the massive devastation inflicted by the Black Death. The plague was caused by infection with bacteria called *Yersinia pestis*, which was transmitted when one was bitten by flea carried by rats. Historically, there were three epidemic outbreaks of the Black Death, and in each of those, many people lost their lives, providing a major impetus for social change by shaking up the socioeconomic structure. The plague usually began with an abscess that filled the lymph glands with pus before developing into septicemia or causing pneumonia. Once pneumonia developed, it killed almost all the people infected with it and spread directly among the people.

The plague, which is believed to have begun in Ethiopia or Central Asia in AD 541, spread rapidly across Africa, Persia, and Western Europe as well as to the Roman Empire via the rats that moved with the people along the trade routes. This plague came to Constantinople in AD 542. and infected Emperor Justinian of the Byzantine Empire, which is why it was later referred to as "Justinian's Black Death." When the Black Death was in full force in Constantinople, some 5000 people died a day, killing nearly a quarter of the eastern Mediterranean population between AD 542 and 546. The Black Death prevailed for almost 300 years, moving from one part to other parts of the empire and devastating the area in its wake. In this way, Europe finally entered the era of full-fledged epidemics. Everywhere the Black Death passed, there was a huge death toll, trade was paralyzed, and taxes could not be levied, resulting in great economic crisis. As it became increasingly difficult to maintain the military, the social hierarchy also collapsed, worsening the political disorder.[25]

The food production system was hit hard with this chaotic situation, and the agricultural system, which was the basis of the economy, was reorganized as a self-centered manor system, setting the wheel of medieval Europe in motion.[26] As there was no scientific understanding of the epidemic, people had no choice but to become more dependent on religious authority, pushing priests to take on the role of healing illnesses, and science to be controlled by religion.[27] Decentralized political and social systems and the religious authority based on the manor system had since dominated medieval Europe, without significant changes or advancement until the 13th century. In this way, disease occurrence and treatment entered the realm of the gods again in the wake of the outbreak of the Black Death in Europe, and most of the medical tradition as a science was carried on only in the Middle East for a long time.

The Black Death restarted in China in the 1330s after a prolonged recess, and then spread immediately to the European cities and rural areas by firstly entering Italia via the trade ships. Between 1346 and 1352, when the Black Death was in full rage, 25 million people died in Europe, or about a quarter of Europe's entire population. This disrupted the commerce and trades and broke down the governmental functions. The roads were filled with dead people and livestock were roaming around. Since then, the plague had persisted for almost 200 years, causing many parts of Europe to stagnate or retreat both economically and demographically. The people who could not know the cause or treatment of the Black Death perceived it as God's punishment for sin and immorality or believed that it was caused by the poison emitted from infected carcasses or patients. Also, the physicians failed to do more than give the advice of restraining one's activities or practicing abstinence. Many Jews were even sacrificed after they were suspected of having spread the poison.

Those who could not cope with the plague in any way, however, experienced anger and frustration in the end, which again morphed into a denial of religious or political authority. As Europe's population declined due to the Black Death, the supply of food became inadequate owing to the lack of labor required for farming and the drop in agricultural productivity. On the other hand, as the population decreased, the rule of the feudal social system became loose, and the farmers were gradually freed. Especially in the wake of the Centennial War that started in 1337, feudalism based on serfdom labor was further weakened as many serfs participated in the war. The denial of the religious and political authority in the feudal society in the wake of such changes shifted the power from one centered on the lords to one centered on the kings, providing the conditions conducive for changing the economic system centered on the manor. The manor system, which was created by the Black Death, again came to grapple with the danger of dissolution due to the Black Death. In the end, as the feudal system fell apart and the power of the kings was strengthened, Europe started to change in earnest. In particular, the historical experience with epidemics raised the need for a powerful system, which again strengthened the legitimacy of establishing nation-states.

With the strengthening of the power of the kings, the European nations opened the Era of Exploration, in which they went abroad to accumulate wealth and to expand their territory. On the other hand, it also played a role in spreading the infectious diseases of the Old World to the New World. The Europeans were not very good in fighting on horseback, but they had excellent skill in navigating into the far seas. While the Europeans expanded their forces into the Pacific through the Indian Ocean with armed warships, the epidemic that was prevalent in Europe spread gradually to Asia, and to Polynesia through Asia. Afterward, the Black Death reappeared in China in 1855, spreading to other parts of the world, including India, Australia, and Africa, along the trade routes, but there was no longer such virulence and mortality comparable to those observed in the past two large-scale epidemics of the disease, suggesting that natural selection had occurred in the direction of reducing the virulence of the pathogens.

Influence of epidemics in the history of the west and the east

Efforts to fight the Black Death epidemics with God's mercy did not play a single role when the plague was spreading in the 14th century. The Europeans thus began to doubt the divine power and the system of power conferred from God, which triggered the Renaissance and the Reformation. They finally turned their eyes from God and began to look back at themselves. Giovanni Boccaccio, born in Italy, wrote *Decameron* in 1358, a book of 100 stories told by seven women and three men who had spent 10 days in a rural village to avoid the Black Death in Florence. *Decameron* was the harbinger of the beginning of the Renaissance, revealing that the epidemic of the plague and the loss of the absolute authority of God and of the Roman Catholic Church transpired at the same time, thereby symbolizing that the Renaissance was greatly affected by the Black Death.

The Renaissance brought about a great leap forward in earnest in the European civilization, as if trapped water was set free. The Europeans began to think more deeply about human existence, and science and medicine, having escaped them from the divine domain, and also began to evolve the way of thinking based on empirical observations. They began to think that the plague was not really a punishment from God, but a disease with a specific cause. Now, they reached a conclusion that an epidemic is no longer a fearsome disease but a disease that could be treated and conquered.

Michele Foucault explained in his book *History of Insanity* that the power of domination has shifted from visible violence to judicial power, and then to invisible scientific power, since the 17th century. The insane should not be restricted by judicial power but should be made the subject of isolation and treatment by a physician. This change in the European awareness of mental disease implies a significant change in the era. In medieval Europe, where it was believed that the angels and demons created by God wield a profound influence on the body and soul, mental illness was thought to be caused by the influence of the devil. The Europeans, however, had woken up to the fact that mental illness does not belong to the divine domain but is simply one of the diseases acquired by humans. This change in the perception of disease was largely because people turned increasingly skeptical of the existence and roles of God while enduring the violent spread of the epidemic, making them question the legitimacy of the Church or the powers that purported to speak for God. As such, the epidemic of infectious diseases provided the Europeans an important opportunity to rethink what was believed to have been granted by God.

Meanwhile, in China, people began to focus on stability rather than innovation after the beginning of the Ming Dynasty in the 14th century, while science no longer developed but stagnated, and the exploration and exchanges with the outside world were suspended abruptly ever since Zheng He led a fleet of over 300 ships and sailed to Africa seven times across the Indian Ocean a century ahead of Christopher Columbus's discovery of America. Whereas Europe underwent changes and achieved developments in philosophy, art, politics, and social structure following the Renaissance in the 14–16th centuries while enduring the severe epidemic, East Asia and China remained relatively stable and conservative societies based on bureaucracy. There was some unrest, such as the rebellion of the peasants against the established social order, the threat of the northern tribes, and the change in the dominant political powers, but it was not up to a level enough to fundamentally dismantle the existing social structure and system. One of the reasons that the socioeconomic infrastructure and political order in East Asia remained largely unchanged was that there was no large-scale outbreak of an epidemic unlike in Europe. In other words, East Asia, another center of the Old World, did not experience the fear of death stemming from large-scale infectious disease epidemics, which is one of the reasons that it managed to maintain a stable society for a long time.

Maintaining stability, however, is also the process of losing the drive for innovation over time. As Europe underwent a massive epidemic of the Black Death in the 14th century, the feudal system was dismantled, and the religious beliefs that sustained the society were shattered, highlighting the need for new thinking and a new order. Propelled by such demands, they secured the drive for social development by strengthening the state power, exploring new colonies, and accumulating wealth. Moreover, based on such drive, science and technology continued to develop, solidifying the dominant position of Europe that lasted for centuries. Europe finally became a model of political, social, cultural, and economic standards while maintaining a leadership in science and technology. As East Asia had never been overwhelmed by a massive infectious disease epidemic, the foundation of its society had not been shaken, which was one of the key reasons that East Asia had been behind Europe in terms of development ever since in nearly every field, with its medical field being particularly far behind that of Europe. To recap, one of the reasons for the relatively stagnant development in East Asia then was its lack of experience of a terrible epidemic like the Black Death, whereas Europe which experienced the Black Death made a major leap forward in nearly every field, including science, from the 16th to the 20th centuries.

Medicine equipped with the potential to develop into a science

The development of medicine remained almost stagnant in medieval Europe until the end of the epidemic. Most people accepted the disease as the will of God, and saw the epidemic of plagues as signifying that the end of the world was near. The physicians did not receive adequate medical education, and the medical knowledge and practice had not been developed much ever since it started in Greece; nor could they be passed on properly to the following generations. Actually, Greek medicine was not succeeded and developed in Western Europe but in Eastern Europe, centered on Constantinople, and even after the Byzantine Empire was overthrown after Islamic domination, Greek medicine still had a considerable influence on the Muslim culture and developed further there. For example, Avicenna's *The Canon of Medicine*, an encyclopedia containing medical theories and knowledge on surgery, medication, and hygiene, had long been used as a medical textbook in Islamic cultures. In fact, plenty of Islamic medicine had been translated into Latin, laying the foundation for the advancement of European medicine.[15] In other words, it is not that Greek medicine went straight back to Europe but that it was further developed first in Persia before returning to stagnant Europe.

By the 11th century, patients had already been categorized according to their diseases in the Islamic world, and hospital facilities had been established to differentiate patients having similar symptoms and signs from other patients, and they taught medicine to students in hospitals for practical training. Education facilities for teaching medicine had only started to spring up in Europe after

Islamic medicine and its practical skills came to the European continent, which then merged with educational facilities for teaching theology and law, developing into full-fledged universities. Medical schools, which had established themselves as educational facilities in the universities, were now the center of medical education. Medical education until the end of the 18th century, however, was not intended to train students in clinics or hospitals but to teach the contents of the medical textbooks that were based on Galenic medicine or on Islamic medicine, which inherited the former.

As Galenic medicine was fundamentally based on the four-humors theory, it assumed that people will remain healthy as long as the balance of humors is sustained, or when a specific humor is excessive or insufficient due to the broken balance, it suggested that the body be cured by extracting an excessive humor by bloodletting, or eating herbal medicine to supplement the insufficient humor. As it was difficult to see such effects in the epidemics, however, Galenus and the four-humors theory gradually lost their authority. In particular, a scholar named Paracelsus declared the end of Galenic medicine. A medical professor at the University of Basel, Switzerland, Paracelsus burned the books of Hippocrates and Galenus in front of his students in a symbolic gesture of breaking with the traditions of the past and looking for new ones. Galenic medicine, however, was not brought to an end by Paracelsus but with the development of anatomy and physiology.

The first human anatomy dissection was performed in 1315 by Mondino de Luzzi of Bologna. As the practice of human anatomy dissection increased gradually, the human anatomy became the central content of medical education, and the *On the Fabric of the Human Body* published in 1543 by Andreas Vesalius of Belgium presented the anatomical structure of the human body in detail. Only after the publication of this book did it become possible to correct Galenus' anatomical error that blood moves from the right ventricle to the left ventricle. In 1628, William Harvey observed that when the heart contracts, blood is pushed out of the heart, and that when the heart relaxes, blood flows back to it. After that, he established the concept of blood circulation based on the anatomical knowledge, which led to the development of physiology, thereby driving medical science to completely depart from the influence of Galenus, and to establish a new medical system. At last, medical science was ready to develop into a legitimate science.

1.4 Biomedical View of Diseases as the Basis of Modern Medicine

The Industrial Revolution was an optimal condition for infectious diseases to spread, and provided a foundation for tuberculosis and cholera to make their presence in earnest in the history of epidemics. Meanwhile, the modern medical system was developed as the activity of clinics, the classification of diseases, and the idea of pathogenesis were gradually integrated as a unified system. In addition, the idea of etiology, which assumes the specific cause of a specific disease, was completed in Europe. Because East Asia was not able to respond well to the epidemics in the 19th century, it was forced to accept the European germ theory and epidemic management system. This again became the decisive moment for Western medicine to rise to the center of medicine in place of the traditional Oriental medicine in East Asia.

The industrial revolution spreads tuberculosis and cholera

The 18th-century Industrial Revolution was a historical event in which urbanization and industrialization progressed at a rapid rate as the accumulated socioeconomic changes and scientific developments began to affect the overall aspects of the society in earnest. In England, efforts were made to establish a new system of the modern state, and the transition from farming to ranching occurred extensively with the dismantling of the manor system, the basis of feudalism. With the decimation of the population due to the Black Death, the supply of labor became scarce, whereas the available land increased, thereby prompting efforts to increase labor productivity. As ranching is possible only with a few people who care for livestock and dogs, whereas farming requires the efforts of many farmers, a considerable amount of farmlands were converted to lots for ranching. The increasing consumption of beef and milk, which was prompted by increasing wealth, further accelerated the transition from farming to ranching. As the need for farmers decreased in the rural areas because of such transition, many farmers had to leave the rural areas and migrate to the cities. Finally, the farmers migrated en masse to the cities and settled down there as members of the lower classes, thereby providing the basis for supplying the cheap labor required for the Industrial Revolution.

Great Britain, which rose to become a colossal empire with many colonies during the Era of Exploration, imported a massive amount of cotton from India. Easier to wash than woolen fabrics, available in a wide range of colored products, and comparably less expensive than other fabrics, cotton fabrics were popular among the middle class in the 18th century. As the demand for cotton fabrics grew significantly, Britain began to manufacture spinning machines to compete with the cotton fabrics imported from India, and after several iterations of improvements, the machine became capable of mass-producing fabrics at a scale much larger than the domestic production. In addition, the steam engine powered by coal provided a turning point for the manufacturing process from one dependent on the human hands to another using machine, resulting in the emergence of a new labor form in which a large number of workers operate the machines in a large factory.

The urban lower classes became factory workers while living clustered in poor residential areas. Living on low wages, they could not afford good food; nor could they obtain clean water because the city infrastructure was not properly equipped to provide such, and the sewage and waste were not treated appropriately. The Industrial Revolution greatly contributed to economic development and modernization but came at a price: the deterioration of the living standards and the urban sanitary conditions for the lower-class workers. This was an optimal condition for infectious diseases to spread, and provided a foundation for tuberculosis and cholera to make their presence in earnest in the history of epidemics following the outbreak of smallpox and Black Death.

Tuberculosis had been around since the beginning of civilization, but until the Industrial Revolution, it had never occurred on a large scale because it was not a rapidly spreading epidemic disease. As industrialization and urbanization proceeded in the wake of the Industrial Revolution, however, the hygienic conditions in urban areas deteriorated to a level worse than before while the living and working conditions became poor, setting the perfect condition for the tuberculosis bacterium to prosper. From the 19th to the early 20th centuries, in particular, tuberculosis spread rampantly among the urban poor populations, with tuberculosis-related deaths in the United Kingdom, accounting for about 25% of all the deaths in the early 19th century. Tuberculosis is often referred to as "consumption" in the sense that it consumes the entire body with chronic progression of the disease characterized by cough, bloody sputum, fever, cold sweat, and weight loss, which combined to weaken the body to death. In most cases, tuberculosis develops in the lungs, but it sometimes spreads to the other organs as the disease progresses. It is also a contagious disease as it spreads to other people when the patients are coughing or through their sputum, making it easy to spread particularly in an environment where the hygiene is poor and in densely populated areas. It was not only Great Britain that suffered from an epidemic of tuberculosis at the beginning of industrialization. The countries that underwent industrialization almost always faced the same fate as Great Britain in the 19th century.

Cholera is a disease caused by *Vibrio cholerae* infection in the small intestine. Patients suffer from profuse and watery diarrhea that lasts for several days, and in severe cases, profound dehydration and electrolyte imbalance occur within just a few hours after infection. Severe dehydration can cause the eyes to sink deep into the eye sockets, and the skin and limbs to lose their elasticity and become wrinkled. The patients can die within a few days if the dehydration is not corrected. In Europe in the 19th century, cholera was an epidemic feared most by the people as it spread quickly and killed about 50% of the infected people. Until the cause of cholera became known, the people thought that the disease was caused either by dirty air called "miasma" or by bad odor.

It was only in 1854 that John Snow of England revealed that drinking contaminated water is the cause of cholera, and it was Robert Koch of Germany who succeeded in identifying the cholera bacteria for the first time in 1884, confirming the *V. cholerae* as the direct cause of cholera.[28] In fact, cholera was a symbolic disease in the transition from premodern medicine to modern medicine because the cholera bacteria had been established as the cause of the disease through three phases: (1) miasma was singled out as the main culprit before the establishment of modern medicine; (2) contaminated water was confirmed as the cause of the disease in John Snow's well-designed epidemiological analysis; and (3) Koch's scientific research revealed that *V. cholerae* is the bacteria responsible for the outbreak of cholera.

Emergence of modern medicine, which assumes the specific cause of a specific disease

It can be said that modern medicine began to emerge near the end of the 18th century. As evidence-based empiricism became the dominant philosophical and scientific trend in the era, only specific actions like surgery and experiments were considered objective and empirical, and medicine without objectivity or verifiability was squeezed out over time from the center of medicine. Doctors began to look into the human body by observing the body structure through anatomical dissection and by further observing the cells that make up the body with a microscope. Medicine became a full-fledged science by looking at the human body objectively and empirically, and the past thoughts and theories on disease were denounced as nonscientific. As such, "the biomedical model" with the mechanistic concept, which assumes that a disease occurs when an abnormal phenomenon transpires in an objective and empirical way in a specific organ of the body, was established as the central theory of medicine. In fact, the biomedical model came from the historical background in which "positivism" had a dominant influence not only on philosophy and science but also on every aspect of everyday life. Disease occurrence and treatment were explained and implemented based on the belief that any disease certainly has a cause and the corresponding effect, and that this should be proved through empirical means.

The empirical observation of a disease in search of its cause and the effort to resolve the cause, however, is not so much an approach that sees the patient as a human being living everyday life but as a target of medical observation. Therefore, the patient was not recognized as a social entity but as a body separated from the society and then occupied by a disease. In addition, disease was understood as a physical state where the inherent body structure or function was damaged when a certain external factor exerted an influence on the body or on the organs, which were made up individually and segmentally. For example, hepatitis B occurs when the hepatitis B virus infiltrates the patient's body and then causes inflammation in the liver, thereby impairing the patient's normal liver function. Therefore, the widely accepted treatment of hepatitis B is either to block the activity of the hepatitis B virus or to take a number of measures to protect the liver functions. This biomedical model occupied the center of Western medicine as scientific rationalism dominated the world since the end of the 18th century, and eventually became the basis for achieving remarkable performance in treating patients compared with the previous one.

Giovanni Morgani of Italy, who greatly contributed to the establishment of the biomedical model, introduced examples of hundreds of patients and catalogued such patients' autopsy findings in his book *On the Seats and Causes of Disease* published in 1761. He recorded the pathologic findings of diseases from each patient, and explained disease in detail as a pathological phenomenon occurring in certain organs and tissues. In addition, he attempted to better understand disease and improve the diagnostic capability by closely observing the patient during illness and performing an autopsy after the patient's death to compare the clinical and autopsy findings. To sum up, he achieved a remarkable breakthrough in medicine by understanding how disease progresses based on the pathological findings from autopsy. Medicine finally left the grip of Hippocrates and Galenus once and for all and succeeded in establishing itself as a completely new medical science. Medicine now became dominated by the new perception that disease is not the problem of the entire body caused by the interrupted harmony and balance of the body but the result of the pathological process of each organ or tissue that makes up the body.

In fact, the classification criteria of the diseases themselves were not clear at all until the middle of the 18th century. Only after the end of the 18th century were physicians able to define the disease appropriately by collecting data from patients and observing their clinical progression. Diseases were now established as objective entities after they began to be classified and named. The biomedical view of disease, which explains how certain factors give rise to a disease in the human body from a viewpoint of pathogenesis, and therefore assumes the elimination of the causal factors of a disease as its cure, was finally established. In fact, such a view of disease excellently explains the infectious disease epidemics that have had a profound impact on the lives of many people even to this day, the reason being that infectious diseases have now been fully explained as being caused by certain germs entering the human body and attacking the bodily organs to cause and propagate diseases. Especially, ever since Pasteur and Koch discovered bacteria, which were the root causes of infectious disease epidemics, such view of diseases has been established as the central theory of medicine. To conclude, the modern medical system based on biomedical models was formed and developed as the activity of clinics, the classification of diseases, and the idea of pathogenesis were gradually integrated as a unified system ever since the end of the 18th century.

Disease-centered instead of patient-centered medicine

The modern medical education and clinical care that began to develop centered in Paris near the end of the 18th century are worthy of being called a "medical revolution" in the sense that they were developed to a far more advanced level than the previous medical education and clinical care. This change can also be seen as a response to the revolutionary changes that transpired in politics and society. Paris at the time was the hub of the medical revolution as well as of the political and social revolution. As was the case with the French Revolution, however, the medical revolution did not go smoothly. Along with the collapse of all the systems denounced as composing the *Ancien Regime* in the course of the French Revolution, the old medical systems, such as the medical schools, hospitals, and physicians, also collapsed. It soon became evident, however, that the revolutionary leaders' idea that a healthy society without disease would be realized once the past privileges and inequalities were abolished was so naive. Disease did not disappear at all, and the soldiers who had participated in the revolution began to ask physicians, especially those physicians who were equipped not only with medical knowledge but also with the capability to perform clinical treatment. In the end, the revolutionary leadership had to create a new medical system, and the technology and education of clinical medicine began to develop centered on hospitals in response to such demands.[15]

As patients gathered at the hospitals and clinical examination techniques were developed, four types of physical examination technique (i.e., inspection, palpation, auscultation, and percussion) began to be standardized and adopted universally. Physicians were now able to examine the patient at a closer distance, unlike in the previous medical practice, when they observed patients from a distance. In particular, René Laennec devised a stethoscope with bell- and plate-shaped ends using a hollow wooden rod, which was very convenient for use in listening to the sound inside the patient's body. Laennec diagnosed many diseases in the heart and lungs by listening to the sounds using the stethoscope which to date remains the symbol of medical instruments. As he was particularly interested in diagnosing pulmonary tuberculosis with the sound of the stethoscope, he examined pulmonary tuberculosis patients enthusiastically. Unfortunately, he died from pulmonary tuberculosis after being infected by a patient.

As such, the physicians who were passionate about treating patients made Paris the center of clinical medicine. In fact, the reason that the knowledge of pathology and diagnostic techniques developed significantly in Paris was that the medical care shifted from house calls to the wealthy class to hospital-centered care in the wake of the political and social revolution. Many patients were hospitalized, with the vast majority of them being soldiers or poor and underprivileged people. As the patients were concentrated in the hospitals, the physicians were able to access them much more easily, and were able to perform tests and develop various diagnostic techniques. Moreover, as it was easy to compare the physical examination findings with the findings from post-mortem autopsies, medical knowledge and technology was able to develop remarkably.

In the meantime, many medical students who came to Paris from all over Europe and from the United States returned to their respective countries to establish medical schools and affiliated hospitals, thereby spreading clinical medical knowledge and technology to many countries. The clinical medicine that developed around Paris as such saw further advancement with the systematic addition of research experiments to clinical examination in Germany beginning in the mid-19th century. This led to the development of cytopathology and bacteriology centered on Germany, which played an important role in perfecting the theory on etiologies of diseases, thereby completing the biomedical model combining clinical medicine and basic medical science. In other words, the idea of etiology, which assumes that an illness occurs when external factors such as bacteria enter the human body and change the tissue cells of a specific organ to abnormal status, or that the cause and effect are connected one-to-one, was completed. Such idea of etiology has since been rooted deeply as the central idea of modern medicine. Such a biomedical model, however, cannot be deemed a patient-centered model but a medical model centered on disease. Accordingly, modern medicine has since practiced disease-centered medical education, research, and treatment rather than patient-centered ones.

Western medicine replaces traditional oriental medicine in the east

The key event in the 19th century in world history was that India and China were incorporated into a single system of the world led by Europe. India prospered during the reign of the Mughal Empire in the 16th and 17th centuries, and was considerably richer than any other country else in the world at the beginning of the 18th century. Its agrarian economy was very productive, and its manufacturing economy was expanding as well. Meanwhile, China shifted from the Ming Dynasty to the Qing Dynasty in the 17th century, and during the period of Qing Dynasty, it was ruled not by the Han Chinese but by the Manchurians. A multicultural regime from its inception, the Qing Dynasty was therefore in need of an absolute power to maintain the regime. For this reason, the Qing Dynasty, with its powerful centralized system, pushed for imperialist expansion based on well-armed troops. In addition, to strengthening its dominance in the central Eurasian area, it expanded its forces further down to Southeast Asia and invaded Myanmar and Vietnam, creating tributary relations there. Its military expansionism, however, largely turned out to be failure while the irregularity of its foreign policy led to the decline of national power in the late Qing Dynasty, along with the corruption by the royal family. Notwithstanding of this, China was one of the world's leaders in the 18th century, as measured by its wealth and the standard of living of its people. The increase of rice production and the introduction of new crops like corns and potatoes resulted in improvement of the people's living standards and overall population growth. The trade with other countries expanded as well, and tea, silk, and manufactured goods flowed from China to other regions of the world. Silver, which was mined mainly in the Americas, flowed into China as a payment for the Chinese exports, helping the country amass considerable wealth.[29]

Therefore, the 17th and 18th centuries were the period when India and China became imperial powers and played important roles in shaping the world economy through active exchanges with the outside world.[30] At the same time, the two countries accounted for much of the global production and consumption then. Nonetheless, Europe in the 17th and 18th centuries had advantages over India and China in terms of imperial expansion as it inherited the legacy passed down from the Era of Exploration. In particular, Europe had a decisive advantage in shipbuilding, navigation, and military forces, the reason being that it had undergone the Renaissance, socioeconomic transformation, the flourishing development of science, and the Industrial Revolution and had successfully created a powerful centralized modern state, since the 14th century.

In the 17th and 18th centuries, India and China were also equipped with an imperialist system based on their military power. There was not a revolutionary change in the countries, however, facilitated by reinterpreting the fundamental roots of their society, such as in Europe, and renewing their worldviews. As a result, India and China were defeated in its competition with Europe, and succumbed to the dominant influence of Europe, which had already seized the upper hand. On the other hand, this provided an important opportunity for the two countries to get connected to the world in earnest. Moreover, the system interlinked as one world has become the base for epidemics to spread globally even if they break out in any one area.

Cholera was the typical disease among the pandemics that spread beyond the confinement of local epidemics. It was first discovered in India in the 11th century, but it did not break out of India for a long time. By 1817, however, it began to spread outside India across the world, including Great Britain and Europe as well as to East Asia, including China and Korea. Since then, there had been six pandemic outbreaks in the 19th century alone. Unlike the previous epidemics, cholera affected not just one continent but the whole world. As the Black Death originated in China in the 14th century before spreading along the Silk Road and sweeping the European states, cholera originated in India in the 19th century and then spread globally through European imperialism, resulting in massive deaths all over the world.

On the other hand, Europe and Asia showed considerable differences in their countermeasures against cholera in the 19th century. Such differences stemmed from their scientific knowledge as well as from their socioeconomic differences. For example, hygiene and isolation were key policies based on the germ theory in Europe, whereas in East Asia, the main policy was to resolve the public angst through ritual and relief. Therefore, whereas Europe was able to respond relatively well to the epidemics in the 19th century, East Asia was not, the reason being that Europe had developed social systems, science, and medicine in the course of fighting epidemics of infectious diseases for centuries, whereas comparable medical concepts had not developed in East Asia then as it had not had much epidemic experiences. Ultimately, East Asia was forced to accept the European germ theory and epidemic management system, which were far ahead of East Asia in terms of theory and practice. This again became the decisive moment for Western medicine to rise to the center of medicine in place of the traditional Oriental medicine in East Asia.

Chapter 2

The Age of Chronic and Late Chronic Diseases: A New View of Diseases

2.1 Humankind Enters the Age of Chronic Diseases

Infectious disease epidemics would soon disappear into history, spurred by the improvement of the public health and the development of vaccines and antibiotics. On the other hand, chronic diseases will dominate because they occur due to the relevant gene's maladjustment to the modern living environment. Therefore, we should pay more attention to the incompatibilities between the genes and the living environments and, more importantly, to the impact of changes in the living environment. However, specific chronic disease is not caused by a specific environmental factor but occurs when the systems of the human body are affected by exposure to specific environmental factors and work beyond their normal ranges. Therefore, the concept of treating a patient with a chronic disease by simply eliminating the cause of the disease has some fundamental limitations. Now, the medical practice must be changed from disease-centered medicine to patient- or human-centered medicine.

The age of epidemics finally draws to an end

In the late 19th century, the incidence of and mortality from tuberculosis began to level off. In the United States, the death rate from tuberculosis reached 194 per 100,000 in 1900, but it dropped to less than 46 per 100,000 in 1944 before streptomycin, an antibiotic for tuberculosis, was used widely.[31] The drop in the mortality rate from tuberculosis was largely due to the practice of hygiene such as patient isolation as well as the improvement of the living environment, such as better diets and housing, which reduced the conditions conducive for the spread of the tubercle bacillus germs. In fact, tuberculosis is a disease that can be contained if the unsanitary environments conducive for the spread of tuberculosis are remedied. In other words, tuberculosis occurrence can be reduced if the probability of bacterial infection in the lungs is lowered and the immunity is strengthened through better nutritional intake, which also increases the possibility of recovery from tuberculosis even if one is infected with the disease. It was not just tuberculosis occurrence that was reduced during this period; infant

The Changing Era of Diseases. https://doi.org/10.1016/B978-0-12-816439-6.00002-8

mortality due to diphtheria, scarlet fever, pertussis, and typhoid fever, which had a major impact on the population, also began to drop noticeably thanks to the improved living conditions such as urban public water and sanitation facilities and other hygiene measures.

Even if the public health was improved, however, communicable diseases remained a threat. In addition to improving hygiene, two more modern medicines, vaccines and antibiotics, were needed to get rid of the threat of infectious disease epidemics.[29] Vaccines are biological agents developed to impart immunity from certain diseases. Vaccines against infectious diseases are usually made by using agents that are similar to bacteria or viruses causing infectious diseases, or that are killed or weakened. When such a vaccine comes into the body, the body recognizes it as an external threat, thereby storing a memory of its intrusion as well as generating an immune response to it. Therefore, if the same strain of bacteria or viruses enters the body again, the immune response will immediately be activated and will defeat the germ-causing infectious disease. Ever since Edward Jenner of Britain proved in 1796 that smallpox can be prevented by using the pus of a cow infected with cowpox, various vaccines had been developed for most of the infectious diseases with high mortality, such as poliomyelitis, measles, chickenpox, and influenza, thereby enabling humankind to prevent many infectious disease epidemics. In particular, vaccines have contributed significantly to the increase of life expectancy since the 19th century by reducing the mortality rate of children.

The era in which patients could only wait for their recovery because there was no special treatment for the diseases caused by germs, such as pneumonia, rheumatic fever, and abscess, came to an end with the discovery of penicillin in 1928 by Alexander Fleming. Fleming observed, during an experiment that he was performing to cultivate pathogenic bacteria, that molds were growing in one of the cultivated plates that had been opened by mistake. Strangely, however, the bacteria were not cultured around the molds. Penicillin has since been concocted based on the clue that substances excreted from the molds kill the pathogens. By the end of World War II, it became possible to mass-produce penicillin, thus opening up a new era in which people responded to bacterial infectious diseases with a powerful weapon called "antibiotic."

It has now become clear that the cause of an infectious disease is a germ, which infects an organ and results in the disease. Moreover, the antibiotic capable of treating diseases by killing the germ responsible for the outbreak of the disease was produced. Therefore, humanity was upbeat after the end of World War II over the prospect that infectious diseases caused by pathogens, like tuberculosis, one of the major diseases that have dogged humankind for a long time, would soon disappear into history, spurred by the improvement of the public health and the development of vaccines and antibiotics. Viral epidemics still break out and spread around the world from time to time, but the era when infectious disease epidemics drove humankind into extreme panic has come to an end.

Chronic disease accounts for two-thirds of all deaths in the 21st century

The number of infectious diseases caused by microorganisms, especially communicable diseases, has decreased, but chronic or noncommunicable diseases, such as diabetes mellitus, hypertension, heart disease, and cancer, have continued to increase in number. Until the early 20th century, pneumonia, tuberculosis, and gastroenteritis were the major causes of death worldwide, accounting for one-third of all deaths.[32] In the early 21st century, however, the major causes of death have become heart disease, cancer, and cerebrovascular disease, accounting for two-thirds of all the deaths. Thus, not only the major causes of death but also the proportion of chronic diseases among the causes of death have changed.

Unlike cholera or tuberculosis, a chronic disease is not caused by a single pathogen but by multiple causative agents, and even if a person is exposed to causative agents, it takes a while for the disease to occur. Also, even after the occurrence of a disease, patients can stay alive for an extended period with the disease, without dying immediately or recovering. Such chronic diseases include hypertension, diabetes mellitus, obesity, heart disease, and cancer. Chronic diseases began to appear after the human race entered the age of civilization, as was the case with the infectious disease epidemics, but their causative agents are unlike anything that caused the epidemics. While the infectious diseases were caused by microorganisms, chronic diseases are caused by many factors associated with the living environment. The living environment, however, does not lead to the occurrence of a chronic disease independently from the genetic makeup; rather, a chronic disease occurs when the human genes and the human living environment cannot achieve harmony and adaptation. Therefore, one needs to understand the concept of harmonization and adaptation that was formed over a long period of time to understand why chronic illnesses occur.

Humankind has gone through the process of natural selection from the age of hunter-gatherers until the age of modern humans. In other words, our hominid ancestors, with genes that could better adapt to the habitat of the hunter-gatherers, had been selected naturally in such a way that they survived and spread their offspring, whereas those who failed to adapt well were not able to leave their offspring. Therefore, the vast majority of the genes we currently have are those genes that have been adapted to the living environment of the hunter-gatherers. These genes, however, cannot adapt well to the modern living environments, the reason being that the modern living environment differs greatly from that of the past, especially one in the age of hunter-gatherers, before the advent of civilization.

Compared with the age of hunter-gatherers, in the modern age, the composition of the people's food intake and the amount of consumed calories changed greatly, along with a significant decrease of physical activity, while new lifestyles, such as alcohol drinking and cigarette smoking, took root. Moreover, the people in the modern age are thrown into much more competitive social relations. Due to the aforementioned changes in humankind's living environment, the human genes that were normal or that helped humankind survive in the past have now become more likely to cause disease, resulting in the occurrence of a host of chronic diseases, such as diabetes mellitus, hypertension, and arteriosclerosis. Especially after the end of World War II, the incidence of chronic diseases jumped to unprecedented levels, along with the sharp rise of the standard of living.

For example, *CAPN10*, a gene known to be associated with the development of diabetes mellitus, is involved in the functioning of insulin. Experiments to block calpain-10 in animals have shown that such action causes diabetes mellitus, while in people, the genetic mutation of *CAPN10* increases the risk of diabetes mellitus.[33] These results suggest that the calpain-10 gene affects the development of diabetes mellitus through its action on insulin. The calpain-10 gene has been shown to activate insulin and push glucose into the cells, making it possible for the cells to use glucose as an energy source. Such function of this gene, however, was established by adapting to the age of hunter-gatherers, when humankind occasionally suffered from famine or ate only low-calorie foods even if foods were available within the radius of their dwelling place, unlike today. In other words, it is a gene involved in the use of glucose as an energy source by injecting the glucose into the cells when the concentration of glucose in the blood is not very high or when it is high only intermittently.

A comparison of today's dietary patterns and physical activity levels with those of the hunter-gatherer times, however, will show that, at present, the amount of ingested calories are often much larger than the calories consumed as energy, thereby easily pushing up the concentration of glucose in the blood to a level much higher than that in the age of hunter-gatherers. Therefore, the elevated glucose content in the blood cannot be fully managed by calpain-10 in many cases, raising the incidence of diabetes mellitus. If the function of the gene to insert glucose into the cell is farther away than the normal gene, owing to a genetic mutation in the *CAPN10* gene, the glucose concentration in the blood will rise further. Therefore, arguably, in the case of the contemporary society whose average blood glucose level is already higher compared with that of the hunter-gatherers, calpain-10's function cannot fully process the blood glucose, making people susceptible to diabetes mellitus. In addition, the added burden of genetic mutation in the *CAPN10* genecan further deteriorate the function of inserting glucose into the cell, making people even more susceptible to diabetes mellitus.

Genetic mutation is not the main cause of a chronic disease

Having observed that diabetes mellitus develops in some people but not in others, many scientists have come to believe that there is a specific underlying cause for chronic diseases, just as some microbes cause infectious diseases.

As there is a genetic code behind every biological phenomenon, it is believed that people with chronic illnesses such as diabetes mellitus contract the disease because of specific genetic mutations not present in healthy people. In other words, the theory of disease development due to genetic abnormality assumes that diseases are caused by genetic mutations, rather than the genes themselves. The reason for this assumption is that genetic disorders like sickle cell anemia or cystic fibrosis are caused by mutations in specific genes. If there are certain genetic mutations in patients with chronic diseases, and these mutations are indeed causing the disease, surely, a biomedical model alone is sufficient to explain such chronic diseases. In addition, the ability to diagnose genetic mutations suggests that disease development can be predicted, and eventual cure of a disease can be realized by eliminating the genetic mutations.

As the technology of genetic analysis improved dramatically in the 21st century, medicine seemed to be on the cusp of a remarkable development for the conquest of chronic diseases. Contrary to the expectations of many scholars, however, it was found in recent years that genetic mutation alone can hardly explain the occurrence of chronic diseases. The rosy dream that gene analysis technology will lead to the conquest of disease has been disappearing. In fact, there is a reason why there is apparent lack of a relationship between genetic mutation and development of chronic diseases. It is because genetic mutation itself does not play a major role in such maladaptation between genes and environment when we argue that the chronic diseases of modern people are caused by the maladaptation of their genes to their present living environments. In other words, chronic disease is caused not so much by genetic mutation as by the failure of the genes themselves to adapt to the contemporary living environment because the human genes had already adapted to the past living environment during the hunter-gatherer period. Therefore, it can be said that a chronic disease occurs not due to the genetic mutation, but the relevant gene's maladjustment to or incompatibility with the new living environment. In other words, the function of the gene itself, rather than the mutated gene, does not fit well with the modern people's living environment.

The calpain-10 gene described here worked well in the past living environment, when the blood glucose levels of humankind were not high, but it does not fit today's calorie-rich lifestyle well. Furthermore, the genes involved in this energy intake and metabolism are not just calpain-10 alone, but dozens of different genes working together. It can be better said that dozens of gene complexes as well as the calpain-10 gene have failed to adapt to today's living environment in the case of diabetes mellitus. Therefore, if the calpain-10 gene's function deteriorates due to the gene's mutation, it is, expectedly, more likely to cause diabetes mellitus. However, the influence of the mutation of one gene on the incidence of diabetes mellitus must be relatively small, compared with the effect of the maladaptation of dozens of gene complexes to the modern living environment.

On the other hand, most genetic mutations do not end up with functional differences, even though the genetic codes are changed. Moreover, even if there is a difference in function caused by genetic mutation, the difference is not usually large. The reason is that genetic mutation is just a way of diversifying the codes of a gene and making the gene more versatile so that it can adapt to a constantly changing environment. Maladaptation of the genes stemming from living environmental changes, however, can be understood as a serious maladaptation caused by the failure of the gene complex to the environmental changes. Therefore, environmental factors like dietary habits, exercise, cigarette smoking, and alcohol drinking can often increase the risk of chronic disease by more than 100% as they greatly alter the adaptability of the genes. It is not likely, however, to see the cases that genetic mutation, a fine-tuning device, increases the risk of disease by more than 100%. The genetic variations that have been shown to affect disease mostly account for only about a 20% increase in the risk of chronic diseases. In addition, the combination of various environmental factors often synergistically affect the occurrence of disease so that the risk increases exponentially, but, in the case of genetic mutation, even summing up all the gene mutations that increase the risk of disease rarely increases the risk over 50%.

In fact, fine-tuning devices, such as genetic mutation, have been used for environmental adaptation because some of the genes randomly mutate to better adapt to the environment. In other words, some of these mutations show minor differences in gene function, which provide the basis for the selection of mutations that make the gene better adapt to the given environment. If natural selection for these mutations occurs in a generation, then the proportion of genes that have good adaptability to the environment will increase. As time passed, descendant generations would consist of more and more people with a certain genetic mutation which is advantageous in selection, and then the said genetic mutation will no longer be considered a "mutation" when everybody possesses such mutation. At this stage, the genetic mutation showing a difference in function will no longer be present, and the gene itself will have taken on the function of the genetic mutation to have superior functions to respond to the given environment.

This is not the end of the genetic adaptation process, however; another genetic mutation occurs randomly and repeats the same process, gradually adapting to the ever changing environment. Therefore, most genetic mutations, in the case of mutations affecting genetic function, are basically associated with small differences in the gene function, and, therefore, they just have a slight effect on diseases, particularly on chronic diseases. In the end, we should pay more attention to the incompatibilities between the genes, but not genetic mutations, and the living environments and, more importantly, to the impact of the changes in the living environment.

Mankind's changing living environment causes chronic diseases

Today's living environment is not only quite different from the living environment of our ancestors; our exposure period to our current living environment is also limited to a few hundred years at most. Especially, the environmental exposures in modern society after the Industrial Revolution have an even shorter history. There was no problem of chronic illnesses in the past, when there was no difference between the time required for the living environment to change and the time required for people to adapt to it. As the time that it takes for a new environment to take root has been significantly shortened, however, the time allowed for genetic adaptation is not sufficient, resulting in the incompatibility between the genes and the environment. This incongruity eventually causes modern humans' chronic diseases. Moreover, the pressure of natural selection is no longer effective due to the recent decline in mortality, which transpired with the development of medicine. Therefore, the fact that the incongruity between the new environmental exposures and the genes cannot be resolved through the natural selection process is one of the reasons that we see such a high prevalence of chronic illnesses today.

As mentioned earlier, when a chronic disease occurs due to the relevant gene's failure to adapt to the new living environment, the occurrence of the disease cannot be entirely attributed to the gene. Rather, it is reasonable to consider the profound change in the living environment as the main cause of the chronic disease. As the genes have adapted to the given environment over a long period of time, it makes sense to say that sudden changes in the environment would lead to the genes' maladjustment to the environment and to the eventual occurrence of chronic diseases. First, let's look at humankind's food intake. Humankind's dietary shift from vegetables, fruits, nuts, fish, and wild animal meat during the age of hunter-gatherers to the staple crops-oriented one after the Agricultural Revolution as well as the marked increase of animal fat intake from consumption of livestock meat after the Industrial Revolution can be deemed to have exerted a significant influence on the occurrence of diseases.

The same is true of alcohol consumption and tobacco smoking. Alcohol consumption has been part of the lifestyle of people since the beginning of civilization, whereas tobacco smoking, which the people during the age of hunter-gatherers were not exposed to, spread to all races from the 15th century. Both alcohol drinking and tobacco smoking have significant effects on almost all the organs of the human body, leading to the development of chronic diseases such as heart disease, diabetes mellitus, and hypertension, as well as cancer. Another problem is insufficient physical activity of people today compared with those of hunter-gatherers. A certain amount of physical activity was basically necessary then because our ancestors had to hunt through "endurance running" during the age of the hunter-gatherers, and had to carry the carcasses of heavy animals over long distances. That explains why our hominid ancestors had engaged in greater physical activity, and their genes had been optimized for such physical activities. Therefore, when the amount of physical activity or exercise is insufficient, the human body cannot operate normally, possibly eventually stumbling upon a chronic disease.[1]

The lifestyles of individuals, however, are just among the many factors that cause chronic illness. Factors like environmental pollution and the increased chemical use in daily life can also lead to the development of chronic diseases. As we look at the environmental factors surrounding us, we can easily see that humans began to be exposed to most of them, such as air pollution, food additives, plastics, and chemicals, only recently. These new exposures aggravate the mismatch between the genes and the environment even further, making people more susceptible to chronic diseases.

Age of chronic diseases: from the disease-centered approach to the people-centered approach

Both infectious disease epidemics and chronic diseases have occurred since the beginning of the human civilization, but their respective underlying factors are different. Infectious disease epidemics first occurred when the human race began farming and herding, thereby living close to domestic animals like livestock while at the same time expanding their active radius through frequent movements and exchanges. In other words, infectious disease occurred as they began to be newly exposed to pathogenic germs, along with increasing opportunities to experience new environments, suggesting that specific factors, such as pathogens, had caused the epidemics. Chronic diseases, on the other hand, occurred because the genes that had adapted to the age of hunter-gatherers or other past environments could not adapt yet to the new living environment. In other words, the disharmony between the genes and the living environment, for instance, causes the blood pressure to rise to high levels, hinders the proper use of elevated blood sugar, thickens the artery wall and clogs the blood vessels, or produces cancer cells.

With regard to the risk factors associated with chronic diseases such as hypertension, heart disease, diabetes mellitus, obesity, and cancer, one can single out unhealthy dietary habits, lack of exercise, cigarette smoking, alcohol drinking, or stress. Although the above-mentioned diseases have different diagnostic criteria and different clinical manifestations, they share almost the same risk factors. In other words, the same risk factors cause different diseases, a phenomenon that cannot be explained with the biomedical model based on the mechanistic causation theory, which assumes that a specific disease is caused by a specific factor. In fact, this phenomenon cannot be understood simply by looking at the relationship between the environmental factors and disease occurrence. It can be understood only after figuring out the complex actions (i.e., metabolism, immune reaction, and energy use) that occur inside the human body when it is exposed to such factors, as well as the genetic and epigenetic programs that direct and control such actions. In other words, it is not that a specific chronic disease is caused by a specific environmental factor but rather that it occurs when the complicated systems of the human body are activated upon exposure to complex environmental factors and work beyond their normal ranges. As the action of the human body system may vary from one person to another even when exposed to the same environmental factors, however, it could appear as a variety of different diseases, such as hypertension and diabetes mellitus.

With regard to the causative factors, it can be said that chronic disease occurs due to the gene's inability to adapt to the changed living environment of modern people. It would be reasonable to assume, however, that complex causative agents interlinked with one another like nets, rather than a single factor, exert an influence on the development of diseases as the living environment of modern people is very complicated. It can be inferred, therefore, that we need to adopt a new medical approach that identifies and manages the effect of complex causes interwoven together. The simple management strategy of the environmental factors that cause chronic diseases would not be sufficient in completely overcoming chronic diseases; rather, one needs to grasp the intricately interwoven networks and the working patterns of the systems inside each person's body and help the genes and each system inside the human body work normally. Therefore, patients cannot be managed properly with the preventive and therapeutic methods developed for the average patients. Preventive methods and treatments tailored for each patient, taking into account such patient's specific genes, environment, and lifestyle, should be developed rather than invariably applying the same management plan to all patients.

In particular, the clinical practice in hospitals that plays a central role in patient management at the moment should be changed. At present, disease-centered treatment is being carried out in hospitals, which involves simply managing patients as people who have certain illnesses to be treated. This disease-centered treatment presents a serious problem. For example, more than half of the elderly today have chronic diseases, with many suffering from more than one chronic illness at the same time. As such, in this case, the disease-centered system is not only inefficient but can also cause considerable confusion as different treatment modalities are applied separately to each different disease. In fact, this system is based on the biomedical model, which assumes that different diseases must be treated independently because different specific factors cause correspondingly different specific diseases. Although it can be said that disease-centered medical care contributed greatly to the enhancement of medical expertise, it is not easy to effectively treat patients with the concept of such mechanistic response in the age of chronic diseases.

As the number of aging people is increasing, the boundaries between health and disease have become increasingly unclear. In a disease-centered system, healthy people and patients are managed separately, but confusion arises when this distinction becomes unclear with the growing population of the elderly. It is because aging itself reduces the function of each organ and often results in a body function somewhere between health and disease states. In addition, the concept of treating a patient with a disease by simply eliminating the cause of the disease has some fundamental limitations because the patient mostly had already been exposed to disease-causing factors long before the disease occurred and developed the disease gradually. Therefore, medical care should be carried out based on the life cycle that undergoes growth and changes from the embryonic to the aging stages. In this case, as expected, disease-centered medicine cannot solve the patient's problem sufficiently. In the end, the medical practice must be changed from disease-centered medicine to patient- or human-centered medicine.

2.2 The Age of Late Chronic Diseases Is Looming

After suffering from the epidemics of infectious diseases in the past, humankind managed to escape from them but had to face a widespread epidemic of chronic illnesses. The recent breakthroughs in medical technology, however, have given considerable hope for controlling chronic diseases as well. If the current increase in chronic diseases following the decline of infectious disease epidemics again shifts to a direction of decline, will humankind indeed enter a "disease-free age"? The average life expectancy will probably increase with the decreasing mortality from infectious or chronic diseases, but humanity will face another problem. In fact, new diseases are likely to increase rapidly along with a decline of the chronic illnesses. The newly emerging diseases include neuro-degenerative diseases, immune disturbances, and mental disorders.

Fast-changing aspects of disease

It is now undeniably clear that the incidence and prevalence of chronic diseases are increasing along with the corresponding increase in the elderly population. According to a report by the World Health Organization, 68% of all the human deaths in 2012 were due to chronic diseases like heart disease, cancer, and diabetes mellitus.[34] After increasing to remarkable levels in recent decades, chronic diseases have now reached the level of epidemics, with about half of the adult population suffering from at least one disease, thereby accounting for up to two-thirds of all deaths. As chronic illnesses usually occur with age, the older the world population becomes, the more chronic diseases are likely to occur. In the meantime, the chronic diseases that were prevalent in the industrialized developed countries until the latter half of the 20th century are now spreading like epidemics in developing countries. By 2030, chronic diseases are expected to be the number one cause of death even in less developed countries like those in sub-Saharan Africa. That is, after suffering from the epidemics of infectious diseases in the past, humankind managed to escape from them but was greeted with the age of chronic diseases in earnest.

New infectious diseases such as those caused by the Ebola virus, Middle East respiratory syndrome, and Zika virus still pose threats to date, but it is clear that infectious disease epidemics are declining as a whole. The recent disease trends show that the share of infectious diseases like tuberculosis and cholera is declining while the share of diseases caused by chronic diseases like cardio-vascular disease, diabetes mellitus, and hypertension is increasing. Although there are some regional differences, chronic diseases are clearly on the rise as a whole, particularly because chronic diseases are increasing in developing countries.[35] If so, will such trend continue to hold up in the coming years or, just as the infectious disease epidemics continued to decline with the improving hygienic environment before it concluded with the development of vaccines and antibiotics, will the age of chronic diseases end with improved living conditions and the advancement of medicine?

To properly look at the health problems of the modern society and effectively cope with them, it is necessary to accurately understand the trends of today's diseases and their changing patterns. In 1991, the World Bank and the World Health Organization jointly launched a study on the global burden of disease. To quantify the impact of different diseases with a single measurement unit, the disability-adjusted life-years (DALYs), which is the sum of the years lost due to an untimely death and the years of living with disability, was used. This is a very useful way of comparing multiple diseases and ranking the burden of different diseases in a consolidated manner as it considers not only the person's death but also the time of the person's suffering from a disease. According to a research paper published in the 2010 *New England Journal of Medicine*, ischemic heart disease was ranked first in the United States in terms of the disease burden as measured by DALYs, followed by chronic obstructive pulmonary disease (second), back pain (third), bronchial and lung cancer (fourth), and depression (fifth).[36] In other words, heart disease, including ischemic heart diseases like myocardial infarction and angina, is the disease suffered by most Americans in the United States, followed by chronic obstructive pulmonary disease (i.e., chronic bronchitis and emphysema).

The study also examined the global burden of disease by surveying 291 diseases and impairments, as well as 67 disease risk factors. Therefore, it covered almost all the diseases occurring in the world, and their risk factors. As the results from the years 1990 and 2010 have been compared based on the survey results of each country, the data can be deemed to contain a sufficient amount of data for disease trend changes. In 1990, the DALYs measured in 187 countries around the world were 2.497 billion, but it was reduced to 2.482 billion in 2010. Considering that there was a considerable increase in population over the 20-year period (1990–2010), the DALY should have increased by 40% if there were no changes in trends of diseases, but, in reality, it was reduced significantly, suggesting that humankind managed to avoid much pain caused by diseases. This is because the vast majority of infectious diseases as well as those diseases caused by maternal and infant health issues or by nutritional problems have been greatly reduced, although the burden of diseases like HIV infection/AIDS has increased. This change is perhaps due to the combined results of the improved health conditions of the mother and the child, enhanced disease prevention and treatment practices, increased utilization of medical facilities, and better living standards.

The risk factors causing diseases also changed significantly during the aforementioned period (1990–2010). Childhood underweight was the most serious risk factor in 1990, but it fell to eighth in 2010, reducing the burden of childhood underweight by 60%. On the other hand, the burden of diseases caused by lifestyle and environmental pollution such as obesity, excessive intake of sugar and salt, insufficient intake of whole grains, and heavy metal lead exposure increased by more than 30%. Meanwhile, the share of the burden caused by increasing disability compared with the burden caused by deaths became much larger with the falling mortality rates. The diseases that cause disability if not leading to immediate death, such as musculoskeletal/mental/neurological diseases, diabetes mellitus, and vision loss, were not reduced but rather increased. The falling mortality also has increased the life expectancy. The data from 187 countries around the world show that the life expectancy of a boy increased by 11.1 years from 1970 to 2010 and that the life expectancy of a girl increased by 12.1 years during the same period.[37] Especially in Japan and some other countries, the life expectancy of the children being born at present is predicted be over 80 for both men and women, thereby confirming that the aging of the entire human race is rapidly progressing.

Chronic diseases decrease in developed countries and increase in underdeveloped countries and among the lower classes

The trend in chronic diseases is also changing very rapidly. The MONICA study conducted by the World Health Organization examined the trends in the mortality rate of cardiovascular diseases by observing 37 population groups in 21 countries for 10 years, from the early 1980s. In fact, this study began with the purpose of confirming whether the decline in the mortality rates of the cardiovascular diseases that had been reported in the United States since the 1970s is also occurring in other populations. The analysis of the data from each country showed that the incidences of cardiovascular diseases and the mortality rates were declining in most countries, except for some population groups in developing countries. Of course, most of the countries that participated in the MONICA study did not necessarily represent the entire human race because they were mostly developed countries that could obtain reliable data. Perhaps the less developed countries that did not participate in the study were likely to be experiencing a rapid increase in their respective incidences of cardiovascular diseases, just as the cases of cardiovascular diseases increased in the advanced countries in the past.[38] At least in this study, however, it was confirmed that the incidences of cardiovascular diseases had been reduced in countries with high income levels.

In recent years, the overall incidence of cancer was also reported to be declining in advanced countries like the United States. Especially, the incidences of frequently occurring cancers (i.e., lung cancer, colon cancer, and prostate cancer) are clearly declining.[39] This phenomenon is actually observed in many European countries as well. In the United States, for instance, the air has been getting cleaner of late, cigarette smoking and alcohol drinking have decreased, and physical activity has been increasing. Of course, not all health-related indicators are improving, for instance, obesity continues to increase, but there is a clear trend toward a decline in some chronic diseases, such as cardio-vascular diseases and cancers, which coincides with a tendency of improving living conditions, suggesting that humankind can reduce chronic diseases by understanding the role of lifestyle and environmental factors, and coping with them appropriately. In view of this trend, it is possible to predict that chronic diseases will be reduced to a considerable extent, at least in developed coun-tries, within a short period of time.[40]

It is important to note, however, that the incidences of chronic diseases are increasing rapidly in countries with low income levels and that the measure-ment of the burden of chronic diseases in the entire human race revealed that about 90% of such diseases are occurring in medium- or low-income countries. In fact, as advanced countries enjoyed a relatively long period of transition to a modern society by going through the Industrial Revolution, it is relatively easy to identify the transition from the age of infectious disease epidemics to the age of chronic diseases. It also appears, at present, that chronic disease epidemics have either reached their peak or have been passed by in developed countries. On the other hand, developing countries are undergoing a rapid transition to a modern society often without directly experiencing the stage of industrial revo-lution and a range of diseases usually observed along various social develop-ment stages are occurring at the same time in such countries, making it difficult to manage such diseases effectively.

As such, chronic illnesses do not occur at the same level across all regions and countries but differ according to the socioeconomic and technological development level of the region and country when examined by the cross-sec-tional approach. In developed countries, chronic diseases have peaked, and their management has begun to show significantly positive results. There are still areas in the world, however, where the basic health indicators, such as infant mortality, have yet to improve. Most other countries have problems with mal-nutrition and infectious diseases while also suffering from chronic diseases like cardiovascular disease, diabetes mellitus, obesity, and cancer. In other words, most countries are in a stage of transition from the age of infectious disease epidemics to the age of chronic diseases.

Citizens from countries that have undergone rapid changes recently, there-fore, suffered from nutritional deficiency at birth but began to be exposed to excessive nutrition when they grow up, compared with the inhabitants of the developed countries, where such changes have occurred for a relatively long period of time, or at least for a period of more than 150 years. In this case, a chronic illness is much more likely to occur if excessive nutrition is provided during adulthood due to the epigenetic program designed to consume the avail-able energy as efficiently as possible in preparation for possible famine because of experience of undernutrition during early childhood.[41] In other words, as considerable changes happened in the living environment in these countries in a short period of time, such countries are expected to struggle with a more serious problem of chronic diseases compared with developed countries. Therefore, humanity will not be able to escape the epidemic of chronic diseases, at least for the time being, due to the increasing incidences of chronic diseases in developing countries.

In fact, the difference in such prevalence of disease at a given time has been observed among the upper and lower classes in the past. Prior to the Industrial Revolution, chronic diseases like cardiovascular diseases or diabetes mellitus had already occurred in the ruling or upper class, such as the royal family or nobility, but not in the majority of the people in the lower classes, the main reason being that the farmers who served the ruling class were not able to receive sufficient nutrition, whereas the ruling class consumed excessive nutrition. The lower classes, who accounted for the majority of the members of the society, however, were able to escape the peril of poverty due to increased productivity in the wake of the Industrial Revolution. Therefore, the classes with higher incidences of chronic diseases changed after the arrival of the modern society, where the food became abundant. Whereas the upper classes became less likely to suffer from chronic illnesses than the lower class as they began to manage the factors that had negative effects on their health, the lower class people could not do the same. In other words, as the lower class people were highly exposed to cigarette smoking, alcohol drinking, or other harmful health factors, while eating more low-quality foods containing saturated fatty acids or *trans*-fatty acids, they began suffering from increasing incidences of chronic diseases. Thus, humankind has experienced a widespread epidemic of chronic illness in such a scale that has never been seen before, while also experiencing a change in the pattern of chronic illness that varies by population group.

Can the advancement of medicine end chronic diseases?

Improvements in the causative factors of the living environment, such as poor diet, lack of exercise, cigarette smoking, alcohol drinking, excessive or too little sunlight exposure, and environmental pollution, may significantly reduce the overall incidences of chronic diseases. Just as it was the vaccines and antibiotics, however, that ultimately brought the end of the epidemics in the case of infectious diseases even if infectious diseases had been significantly reduced by improving the hygienic environment, the chronic diseases cannot be resolved simply by improving the factors of the living environment. The recent breakthroughs in medical technology, however, have given considerable hope for controlling chronic diseases. It can be said that the period in which medical technology began to make a significant contribution to the prevention and treatment of chronic diseases was the mid-20th century, after the conclusion of World War II, as the knowledge of biochemistry and molecular biology developed in earnest only after the mid-20th century, along with an explosive growth of chronic diseases such as cardiovascular diseases, diabetes mellitus, and cancer, thereby prompting the remarkable progress of the pharmaceutical industry. Since then, significant advancements have been made in therapeutic technology, prompting the arrival of an era where most chronic diseases can be managed with drugs. In other words, drugs capable of blocking the mechanisms of diseases have been developed through the scientific understanding of the mechanism involving the occurrence and progression of disease in the human body. Diagnostic tools like computed tomography, magnetic resonance imaging (MRI), and ultrasound have also been made more elaborate, making a significant contribution to the detection of disease and the determination of its range and severity.

Let's take a look at a hypothetical case where myocardial infarction occurred due to the blocking of the coronary arteries supplying blood to the heart muscle. Nowadays, in patients with myocardial infarction, the exact site and severity of the cardiovascular blockage are identified by using electrocardiography, echo-cardiography, MRI, and angiography. Moreover, a stent is inserted in the narrowed coronary artery to expand it, while administering antihypertensive drugs and anticoagulants. Thanks to such treatment techniques, patients with myocardial infarction who would otherwise have died or been unable to recover only a few decades ago can now recover within 1–2 weeks and return to their normal social life. Surgical techniques have also developed dramatically to enable the transplantation of key organs like the kidney, heart, liver, and lung. Moreover, laparoscopic or robotic surgeries with minimal skin incision have now largely replaced laparotomy or open thoracic surgery, where the skin is extensively incised. For instance, robotic surgery is a surgery in which a robot's camera and arm are put inside the patient's body after three or four holes are made in the body, and then the robot's arms are controlled while the surgeons are watching enlarged three-dimensional images on the display. Robotic surgery makes it possible to perform an elaborate surgery because there is no shaking of the operator's hands and the surgery can be performed in a narrow space that can hardly be reached by the human hands.

Pathogenetic mechanisms, although not complete, have also been discovered not only for cardiovascular diseases but also for most chronic diseases, such as hypertension, diabetes mellitus, obesity, asthma, and depression. The development of therapeutic techniques or drugs based on these mechanisms has at least controlled the disease to a certain degree even if the disease is not completely cured. Cancer is relatively difficult to treat at the moment, and the patients are known to have a low survival rate, but remarkable results have also been achieved of late in cancer treatment. In the case of childhood leukemia or female breast cancer, early detection and proper treatment have led to a full recovery and to a healthy life. Although lung and pancreatic cancer patients still show low survival rates, the current pace of medical advancement is expected to eventually lead to a significant increase in their survival rates as well.

The level of medical care has shown definitely a significant improvement over the past. As chronic diseases, however, are inherently caused by the inability of the genes to adapt to the living environment, the chronic diseases that are prevalent today cannot be resolved completely unless the genes, living environments, and maladaptation phenomena are fully understood, and unless such maladaptation state is changed to a harmony and adaptation state. This gene maladaptation, however, is difficult to solve with the concept of simple causality and therapeutic techniques based on the mechanistic biomedical model. Therefore, the next step in modern medicine should be to fully understand the complex nature of maladaptation and to develop advanced treatment techniques based on such understanding.

Another disease emerges following chronic diseases

If the current increase in chronic diseases following the decline of infectious disease epidemics again shifts to a direction of decline, however, will humankind indeed enter a "disease-free age"? The average life expectancy will probably increase with the decreasing incidences of death from infectious or chronic diseases, but humanity will face another problem. In fact, new problems have already appeared, and the new diseases are likely to increase rapidly along with a decrease in the incidences of chronic illnesses, just as the chronic illnesses exploded exponentially after the decline of infectious diseases. The newly emerging diseases include neurodegenerative diseases like Alzheimer and Parkinson disease, immune disturbances such as atopy and Crohn disease caused by disturbed immune function, and mental disorders influenced by increased mental stress.

Diseases such as Alzheimer disease, which triggers dementia that is characterized by memory loss and cognitive impairment, and Parkinson disease, which hinders the movement of the body, occur when the proteins in the brain neurons aggregate together. As the average human lifespan increases, the number of people undergoing the aging process also increases, pushing up the number of people developing protein aggregation in the brain neurons, resulting in neuronal dysfunction. Such increase in the number of people with deteriorating function of the neurons is the reason for the increased incidences of neurodegenerative diseases.

Immune disturbances are caused by the break in the balance and harmony between various factors inside and outside the body, thus disturbing the immune system of the human body. Atopic diseases that trigger abnormally excessive reactions to outside factors, albeit not toxic or irritating, and autoimmune diseases like inflammatory bowel diseases, including Crohn disease that triggers an immune response to the cells normally constituting the body because the ability to differentiate the self from another is impaired, are among the diseases caused by immune disturbance. The underlying reason for the increase in the incidences of mental illnesses like depression is that the excessive competition or stress in the world today increases the body's consumption of neurotransmitters like serotonin or dopamine in response to such stress, which often leads to the failure of the brain's normal response mechanisms.

In fact, these diseases manifest chronic development progress, as is the case with diabetes mellitus, heart disease, hypertension, and cancer, and are often caused by factors that are more complex than single factors. In addition, to the maladjustment of the genes to the environmental changes, which have been the main cause of chronic diseases, another characteristic of such diseases is that new factors which have not been considered causes of diseases, such as aging, changes in the intestinal bacterial flora, and a competitive social structure, are said to help trigger the outbreak of such diseases. These diseases are considered chronic diseases as well, because they go through chronic pathogenetic process, but they can also be called "late chronic diseases" to distinguish them from chronic diseases like diabetes mellitus, hypertension, and heart disease.

The diseases called late chronic illness are expected not to decline but rather to proliferate even after the age of chronic diseases. Whereas chronic diseases are caused by excessive calorie intake, lack of exercise, cigarette smoking, alcohol drinking, and exposure to pollutants, new factors play an important role in the case of the late chronic diseases. In other words, the current change of the human society towards a future society in which the lifespan of the human is greatly increased, the majority of the human settlements are urbanized, and the entire society is connected to the network is itself a fundamental factor causing late chronic diseases. Although chronic diseases or late chronic diseases largely results from maladaptation of genes to living environment, it can be said that personal lifestyle is an important factor in chronic diseases, whereas society change is a much more important factor in late chronic diseases. Therefore, if we go to the future with current development strategy, we will not be able to prevent the epidemic of late chronic diseases.

Now, new strategies are needed to prevent such epidemic. A medical strategy based on a mechanistic biomedical model has already shown considerable problems in the management of chronic diseases. Since late chronic diseases are problems related to society change, we cannot cope with the emerging diseases using such an obsolete strategy based on the model that a specific factor, mostly personal lifestyle one, causes a specific disease. To create a disease-free society in the future, new medical strategies drawn on new pathogenesis models must now be established.

2.3 Disease Occurs When Harmony and Balance of the System Are Breached

It is important to understand the interaction of the various elements involved in the action of the human programs, such as the genes, epigenetic programs, and proteins, to accurately understand how the human body programs work. The same is true for disease development because the various components making up the system of human body are involved in it, and one system affects another system. To validly assess such chain of associations in the relationship between causal factors and disease phenomena, it is necessary to properly link the factors and the disease based on understanding of the pathogenetic phenomenon. This could be accomplished, however, only by deeply understanding the reactive action of the human systems through the information obtained from the data on response of the human systems.

The human body is composed of complex systems

Chronic or late chronic disease is a phenomenon caused by the complicated entanglement between the disease-causing factors and their corresponding diseases. Why, then, is the human body function based on a network of complex systems rather than on a simple one-to-one relationship between specific factors and their corresponding consequences of response? In fact, the human body must have not been complicated in the beginning but only became so as a result of the development of various response mechanisms and the accumulation of such mechanisms over a long period of time to cope with environmental conditions in the course of the evolution of the human body. Ultimately, the human body is a cumulative product of the long-term countermeasures that it had developed against its environmental threats. Therefore, the complex systems of the human body should be examined in relation to the environmental conditions of human. If maintaining the harmony and balance between the systems in the human body and the environment outside it is referred to as healthy status, then a disease occurs when such state of harmony and balance is disrupted. To conclude, disease can be understood as a disrupted state of the human body's complicated network.

Therefore, as humankind can escape the state of disease by restoring the human body's harmony and balance, it is necessary to create environmental conditions that the biological systems, including the human genes, can adapt to in harmony and balance. It is virtually impossible, however, to create an environment well suited to the biological system of modern humans, the reason being that we cannot return to the age of hunter-gatherers or some past period and recreate the living environment of our ancestors in the contemporary world. Of course, we can improve our living habits, such as eating a healthy diet, exercising regularly, avoiding cigarette smoking, and abstaining from drinking alcohol excessively, so that we can live a life that resembles to some extent that of our ancestors whom we inherit the genes or genetic programs from. Nevertheless, we cannot simply go back to the past living environment and ignore the many realistic conditions of the modern society. However, what we can do our best is not to be negligent in improving our environmental conditions and try to restore harmony and balance among the complex networks of our body that has been disrupted in the given circumstances.

As life phenomena, from molecules to cells, tissues, and individual entities, are connected with one another and constitute a complex system, it would be a stretch to assume that we can explain all associated diseases or find a cure for them, although once we have understood specific molecular phenomena at the cellular level through research.[42] If each of the proteins or molecules in our body is given just one role, and if the gene regulating the function of each protein or molecule is specifically defined, the restoration of the network may not be very difficult. The simple relationship, however, in which the cause and the manifestation of the disease are matched one-to-one can hardly be seen in the human body. Take, for example, the cancer-suppressing protein P53. This protein not only inhibits cancer development but also regulates the cellular cycle. In addition, it induces apoptosis (programmed cell death) and is involved in the repair of damaged DNA. A closer look at the protein shows, however, that it does not play several roles at the same time but plays different roles depending on the given surrounding conditions.[43] That is, the same protein, such as P53, plays a variety of roles depending on the given condition.

Suppose that genes, proteins, and certain functions are closely related to one another. We can easily assume that the removal of a gene will result in the protein loss or nonproduction and in the elimination of the corresponding function. On the contrary, we often observe that another gene steps in to maintain the protein production and functions that had been performed by the removed gene, the reason being that the biological system of the human body is connected to each other through the network, and there are various programs that can make up for a certain defect when it occurs. Although not complete, other genes can regain the removed genes' original functions to a certain extent by performing the aforementioned roles instead. Due to this complicated network, our body has considerable resistance and resilience against the impact of the changes in the external environment. On the other hand, once our body's resistance and resilience break down, pushing our body onto the state of disease, it is not easy to turn it back to a healthy state. This is because the complex network itself must be restored instead of merely fixing a specific part of the failed human programs. Therefore, even excellent drugs targeting a certain pathogenetic mechanism of a chronic disease cannot completely cure the disease.

To illustrate the aforementioned point, let's assume that we have a car in front of us. The car's body is interconnected with the engine, transmission, shaft, and wheels, as well as with various other mechanical and electronic devices. In other words, an automobile has a system composed of various parts, as identified earlier. If any part in such system fails, the car will not be able to operate normally. In this case, when you go to the car repair shop, you will see that there is something wrong with a specific part, and if you have it replaced with a new part, the car will run again just like before. If a human body is a system in which a particular gene, protein, or molecule plays a defined role just as every part of a car does, and such things work together closely just as all the parts of a car do, then we should be able to treat a defective or damaged gene, protein, or molecule to cure diseases, just as fixing the damaged car part will enable us to run the car again. The problem, however, is that the human body is different from a car. Whereas a car is a simple assembly system of many parts, the human body is a complex system in which various elements form an organic network.

It is easy to assume that there is a one-to-one mechanistic correspondence among the phenomena occurring in the human body and that the first step is followed by the second step and then by another step and so on, combining to create a single program. This, however, is not what occurs in reality. The programs in the human body are not defined within a certain logical framework but are organically connected with one another and are influenced by time and the given conditions. In short, it is important to understand the interaction of the various elements involved in the action of the human programs, such as the genes, epigenetic programs, proteins, metabolites, mitochondria, and symbiotic microorganisms, to accurately understand how the human programs work.

Approaching disease with a new view

It goes without saying that humans have systems that perform more complex functions than do single-celled organisms. About 4 billion years have passed from the appearance of the first single-celled organisms to the emergence of the human race. During that period, genes, protein synthesis, energy production and consumption, metabolism, and other systems were established, with each system having evolved from a primitive life form to a more complex form. In particular, emergence of eukaryotic cells 2 billion years ago marked a turning point in establishing this complex system, the reason being that the complex system of multicellular organisms must have a nucleus that directs and supervises the complex functions of the cells and that there must be a mitochondrial energy production system that supplies the energy needed to operate under the commands of these nuclei.[44] A cells have been equipped with complex functional systems and efficient energy supply systems, they can evolve into multicellular organisms, plants, and advanced animals with more complex systems, and these changes have been passed down to contemporary humans. Humankind was born with these complex systems, and all the phenomena in the human body are manifested through the operation of such complex systems.

"Complexity" refers to a state in which the components are not simply arranged but are interwoven with other components to play their roles as part of the structure or the function of a system. Therefore, it is necessary to understand the complex network connected to the whole as well as the simple relation among the components to grasp the entire system. On the other hand, the biomedical models built upon the mechanistic causation theory are based on simple linear relationships. For example, the relationship between a pathogen and an infectious disease is understood as a simple linear relationship, such as in $y = ax + b$, and is defined as the proposition that a specific pathogen causes a specific infectious disease. Moreover, the proposition has been perceived as an incontrovertible truth that will not change over time. From the 19th to the 20th centuries, in particular, the biomedical models based on simple linear relationships have dominated medical science, without being challenged. A change, if any, is that a multiple linear model like $y = a_1x1 + a_2x2 + a_3x3 + \ldots + b$, which considers multiple factors simultaneously in a linear relationship, has been newly added along with the recent addition of a nonlinear model. Even if such complements or additions have been realized, however, it cannot be said that the theoretical foundation has deviated from the simple linear relationship.

It is, in fact, an oversimplification to think, however, that biological phenomena occur in such simple linear relationships that do not change over time. On the contrary, there rarely is such a possibility. The same is true for disease because the various components making up the system of human body are involved in it, and one system affects another system, thereby changing such relationship over time. When looking at biological phenomena with assumptions of simple linear relationships, complex relationships that are associated with diseases look randomly irregular, just like background noise, and are difficult to quantify, making it easy to dismiss them as meaningless findings. It is more reasonable, however, to assume that the biological phenomena in the human body transpire as complex entities where various components are influenced by one another in a closely connected relationship.

As the hormones, for instance, change with a rhythm of the biological cycle, they change continuously over time, and such changes in them affect their responses to external stimuli, such as genetic expression and protein production. In other words, the demand for hormones changes every time we eat, move, think, or sleep, and such changes in the demand for hormones also continuously alter the genetic expression and protein production. In this case, the number of affected genes or proteins is not one or two but may add up to tens or hundreds. Ultimately, what happens in the human body is not a simple preset relationship but a complex system that changes over time.

Take, for another example, diabetes mellitus. Caused by a high concentration of glucose in the blood, the disease is commonly known to occur when one ingests excessive carbohydrates. In fact, there have been countless studies that reported the relationship between carbohydrate or calorie intake and the incidence of diabetes mellitus. According to the results of these studies, it can be said that excessive caloric intake is the cause of diabetes mellitus. On the other hand, an analysis of the relationship between exercise and the occurrence of diabetes mellitus, which was performed under the hypothesis that diabetes mellitus occurs because not all the consumed calories were used, revealed that lack of exercise is the cause of diabetes mellitus. Obesity, which is associated with caloric intake and lack of exercise, has also been shown to be the cause of diabetes mellitus. Of course, as dietary intake, exercise, and obesity are all factors related to the intake and use of calories, it can be expected that these results are related to diabetes mellitus.

In recent years, however, studies on the relationship between stress levels and diabetes mellitus have found that stress is one of the main causes of diabetes mellitus, and it was also found that exposure to endocrine-disrupting chemicals, such as dioxins and phthalates, increases the risk of diabetes mellitus. Even fine dust in the air has been shown to contribute to the development of diabetes mellitus. Stress, endocrine-disrupting chemicals, and particulate air pollutants are not related to caloric intake or to one another, and the systems to which these factors belong are not closely related with one another. In other words, factors belonging to different systems are acting as the causes of diabetes mellitus.

These seemingly unrelated systems play a role in causing the common disease named "diabetes mellitus" because they are connected to the system inside the human body. The relevant body systems are, for example, the carbohydrate metabolism and transport system of glucose, which metabolizes food into glucose and transports the sugar in the blood inside the human body; the mitochondrial system that converts the sugar entering the cells to energy; the response or defense systems to stresses and external substances; and the regulatory systems of genes and epigenetic programs governing all the aforementioned systems. Systems that exist independently outside the human body, such as chemical exposure or social relationships, are also connected to and interact with those systems that operate inside the human body. Therefore, external factors, like chemical substances and stress from human relationships, do not act independently but indirectly influence each other. Therefore, diabetes mellitus is not caused by a few limited elements that make up the system but by the disrupted balance and harmony of various systems that affect one another while engaging in action.

Avoid the fallacy of exaggeration and oversimplification

It can be said, therefore, that diseases come about when a variety of systems outside the body, such as microorganisms, living environments, social relations, and temporal processes, activate genes, epigenetic programs, and protein expression, and when the resulting reactions in the human bodies triggered by such systems, including immune, inflammatory, and metabolic reactions, occur beyond the normal ranges. In other words, factors widely scattered across various dimensional areas are involved in the occurrence of diseases.

Even in this multidimensional concept of disease development, however, a more realistic approach for understanding and managing diseases would be to narrow down to a two-dimensional relationship between causal factors and disease outcomes. Although the network of causal relationships is very complex, it is important to figure out the factors that control the outcome of the disease to improve the chances of preventing and treating diseases. This, however, should reflect the fact that the disease is caused by a variety of factors, such as individual lifestyles, various environmental factors, and even social structure and relations, rather than by one or two specific factors, as in the biomedical view of disease.

In fact, if the intricate information coming out of the network of systems is delivered without being streamlined, it is difficult to handle it easily in our brains, the reason being that the amount of information is vast, and not much of its content can be easily understood. Ultimately, complex information needs to be gathered and processed properly with the help of an information system with a processing power way beyond the level of the human brain. Fortunately, recent scientific and technological improvements, including medical, biological, and statistical science advances, provide the basis for processing large amounts of complex information.

On the other hand, the actions people take in response to the way these complex systems are being operated need to be simple in nature. This is because humankind's information processing pattern and their behavior have evolved over time in ways that would facilitate an execution of simple actions by recognizing the patterns of complex information and characterizing them. Therefore, the health management practiced by each person should be made at a level that can make it easy for everyone to understand and perform. In other words, even if we use complex information, it is necessary to develop health behaviors that everyone can easily carry out, and provide practical ways to get disease management without difficulty when going to a hospital if it is medically necessary.

Errors can occur, however, when simplifying complex information. Therefore, it is important to make a very efficient network of causal links by avoiding unnecessary or ancillary factors to effectively prevent and treat diseases, even though one needs to be careful not to miss key elements while in the process of simplifying a complex system. For example, suppose a pregnant woman eats tuna one day, probably pushing up the mercury level a little in her blood, which is then transmitted to the fetus through the placenta. As mercury has a negative effect on the development of the fetus, especially on the development of the fetal nervous system, when the mercury level in the blood is high, the baby may grow at speeds slower than those at which other children grow, and may suffer from attention deficit and hyperactive disorder in childhood. Let's say that this child is not well suited to school life and is later barely able to secure a low-paying job after graduating from high school. This person also leads an unhealthy lifestyle, drinking excessive amounts of alcohol and smoking cigarettes, and has difficulty establishing a good home after marriage. In middle age, he becomes increasingly obese, suffering from hypertension and diabetes mellitus, before eventually dying from myocardial infarction. In this chain of hypothetical scenarios, can the tuna that the pregnant woman ate on one day be assumed to be the cause of her child's myocardial infarction in his adulthood?

One cannot rule out that the tuna ingested by the pregnant woman may be related to the myocardial infarction that occurred to the woman's child in his adulthood, but emphasizing statistical associations too much may lead one to argue for the causality between two separate facts just like "the butterfly effect," arguing that a tornado occurred in Texas due to a butterfly's fluttering of its wings in Brazil. In fact, one butterfly that flaps its wings cannot directly generate tornadoes. The butterfly's wings may be associated with the initial conditions of the tornado development, but they are not directly linked to most of the other conditions that contributed to the eventual development of a tornado. If events unrelated to one another occurred after the appearance of the first event, and these events triggered a larger-scale disturbance, such as a tornado, it will be an overstatement to claim that the first event caused the tornado as the ultimate outcome.

William of Occam, a Franciscan friar who lived in the 14th century, having observed that the debates between medieval philosophers and theologians were not only complicated but also pointless, suggested the introduction of a razor blade to cut out the excessive logical leap or unnecessary premises from a statement.[45] In other words, Occam argued that if something can be explained in a variety of different ways, which among them with the least number of assumptions should be chosen. In other words, when explaining a chain of associations with various hypotheses, the simplest explanation is not only efficient but also devoid of exaggeration and close to the truth. Occam's razor blade has the advantage of eliminating the unnecessary contention and simplifying the existing association of things, thereby clarifying the logic, but sometimes it is difficult to know if the eliminated contention is indeed unnecessary. In fact, the butterfly effect or Occam's razor blade suggests that there is a risk of overextending or underestimating the chain of associations if one does not accurately observe the exact phenomena that are occurring. To validly assess such a chain of associations in the relationship between causal factors and disease phenomena, it is necessary to properly link the factors and the disease based on a deep understanding of the pathogenetic phenomenon, while, at the same time, care should be taken to avoid arriving at either extreme.

Decoding the black box for the identification of the cause

In the latter half of the 20th century, when the age of infectious disease epidemics was over and the incidence of chronic diseases began to increase rapidly, more frequent efforts were made to identify the causes of chronic diseases. The vast majority of such efforts, however, were aimed at identifying the risk factors of disease by examining the relationship between exposure factors and disease outcomes by using the simple relationship model of disease development. This, of course, has to some extent been successful in singling out the risk factors of disease for each chronic disease. For example, the risk factors associated with diabetes mellitus are excessive caloric intake, lack of exercise, obesity, and stress. Although we have identified these risk factors, we could not find a satisfactory basis for why such risk factors cause a disease in a particular person, the reason being that it is not known well how these different risk factors lead to the common disease phenomenon such as diabetes mellitus through the supposedly different mechanisms of action in the human body.

In other words, we have tried to find a simple relation between the disease and what we considered the factors contributing to the disease occurrence without knowing the detailed process of development of the illness. In the absence of knowledge on the pathogenetic process from a causal factor to the onset of a disease, we cannot be certain that the factor actually caused the disease even if statistical results show that a certain factor is related to the occurrence of the disease. This is because it is possible that the third factor, which is related to the factor and the disease at the same time, may have made it seem that the factor and the disease are related with each other. Therefore, only after elaborately investigating the changes that occur in the response mechanisms (e.g., metabolism, detoxification, and immunity) in the body following exposure to certain factors, along with the knowledge on profiles of genome, epigenome, proteins, and metabolites, and after confirming the changes actually leading to the development of a specific disease in the body, can we be certain that a specific factor is the cause of the disease.

Moreover, only after comprehensively understanding the causal linkage based on this pathogenetic mechanism can we be able to prevent a disease by eliminating its cause, or provide precise treatment to cure or prevent the disease from further worsening. If we attempt to understand the relationship by simple association without identifying the detailed process that occurs inside the body from the exposure to a certain factor to the occurrence of the disease, it is as if we are trying to figure out the cause of an airplane crash without knowing the information contained in the airplane's black box.

On the other hand, if we only see detailed changes in the human body without understanding how the disease occurs at a bigger picture, it is difficult to know more than the fact that the human body is just a very complex system where so many changes are occurring. For instance, looking at the metabolomics data of human body, you might feel the stars in the night sky! In other words, it is difficult to know where the beginning of the change is, how it is progressing to a disease, or through what intermediate relationships one ends up having a disease. Therefore, a priori causal theoretical hypothesis based on the reasonable pathogenetic mechanism should first be established to easily understand how diseases occur.

It is almost impossible, however, to start establishing a solid hypothesis from the beginning even if it is based on the hypothesis built upon the reasonable pathogenetic mechanism. It is often the case that the first hypothesis created as such just provides a clue as to how changes occur within the human body. From this initial hypothesis, we can perform further analyses aimed at understanding the causal linkage better. With the further analyses, we can reach a better causal mechanism for the diseases. Repeating this, we can arrive at a complete understanding of the cause of the disease. This could be accomplished, however, only by deeply understanding the reactive action of the human systems through the information obtained from the data on response of the human systems, which provide more precise feedback regarding the hypothesis reiteratively.

2.4 A Step Closer to the Closure of Disease Era

Disease should be understood to be caused owing to the broken balance and harmony between various internal and external factors of the human body. A new medical model for diagnosing and treating diseases based on the concept of the harmony and balance of the whole system can be called the "systems medicine model." Systems medicine is a medical approach aimed at comprehensively understanding and managing the complexity of the relationships between the molecules, cells, and organs that constitute the human body, as well as between humans, microorganisms, and even ecosystems. If humanity advance information processing ability more in the near future, it will be possible to make more accurate diagnoses and to better manage people's health by using the techniques of the systems medicine approach, propelling us a step closer to the conquest of disease.

Identifying the complex systems affecting disease

The human body is influenced by multidimensional systems, but a deeper look at each independent system will show that each of such systems is also very complicated. For example, the environment, lifestyle, and microorganisms are external systems that affect the human body. Let's think about environmental exposure among them. There are hundreds of chemicals that the human body is exposed to every day, either through the air (e.g., fine dust, ozone, volatile organic compounds, polycyclic aromatic hydrocarbons, heavy metals, and hundreds of additional chemicals one is exposed to via smoking or indirect smoking) or through exposure to food or drinking water, or via contact with one's hands or skin. In fact, as the lifestyles of individuals are very complicated and varied (e.g., smoking, drinking, eating habits, exercise, work hours, sitting patterns, sleeping habits), it is not practical or even logical to single out any of them as the primary cause of disease.

In the case of microorganisms, only a fraction of them can be measured with the present knowledge and method. There are 10 times more bacteria living in the human body than all the cells constituting our body, and microorganisms and their host (the human body) are in a symbiotic relationship with each other.[46] The symbiotic relationship mentioned here is a "balanced relationship" for mutual benefit. If this balance is broken, the symbiotic relationship is broken as well, potentially exerting an undesirable effect on each other. In addition, the symbiotic relationship is a type of "adaptive relationship" suggesting that the microorganisms have already adapted to the current lifestyle of a person. As microorganisms have a considerable influence on the formation of defense systems such as the human immune response, if the microorganisms cannot adapt to the newly changed lifestyle of a person, the formation of the defense system will be affected as well, making the person unable to defend himself/herself properly when exposed to the factors potentially causing the disease.

Accordingly, to understand the pathogenetic phenomena of chronic diseases, it is necessary to understand how such complex systems get entangled with one another, and how each individual is being exposed to various external system factors. Perhaps, if we understand the pathways through which the human systems produce complex reactions of the human body when exposed to such external system factors, it will be possible for us to prevent and treat chronic diseases in earnest. In other words, a medical care tailored for each individual can be offered if we understand the defensive mechanism in which each individual human body reacts to external system factors, and use appropriate treatment techniques based on such understanding.

As explained in the previous chapter, when determining the factors that cause a disease, it is desirable to continuously improve the hypothesis on the relationship between the potentially causative factors and the diseases based on the varying level of results from data produced along with the changes in the complexly entangled factors, than to approach them by establishing a hypothesis that a given factor is the cause of a specific disease within a given frame. In other words, it is not that a hypothesis on the relationship between the factors and the diseases is predetermined and does not change further, but that hypotheses of various relations are proposed, and then one that best explains the results is selected as the final hypothesis. In addition, a better hypothesis on the factor–disease relationship is constantly updated through the process of learning and feedback, and prevention and treatment of diseases are implemented based on this knowledge.

This approach can be considered a self-learning disease management that continuously collects and judges information to optimize the health status of the subject, rather than as a routine performance of predetermined disease management methods. These disease management methods include the management of many external factors themselves as well as the prevention and treatment of the target person at the same time, the reason being that the disease is not caused by a specific factor that changes the structure of the human body and deteriorates its function by affecting it, but should be understood as a deviation of the structure and function of the body from the normal ranges owing to the broken balance and harmony between various internal and external factors of the human body.

It is not just a single factor, therefore, that affects the health as previously argued, and each individual factor is influenced by other factors. The entire system cannot be assessed accurately simply by evaluating various factors individually and adding them. Only after assessing the health impacts of individual factors and the interactions of various factors as well as of the entire system interconnected with each factor can we accurately assess the overall impact of such factors on the health. In fact, such assessment was not possible in the past due to technical limitations, because information on physiological, toxicological, and immunological mechanisms as well as on various external factors of the human body should be in place, on top of the information processing capability, to handle such complex body of information. In addition, to accurately assess the exposure to complex external factors, the measurement techniques that have been used so far should be further improved while new technologies need to be continuously developed to ensure better accuracy of assessment.

In fact, these technological innovations have been made and continue to progress at a significant pace. For example, exposure to toxic chemicals such as heavy metals can be determined by measuring the concentrations of certain substances in the blood or urine, or by measuring the concentrations in the hair, not to mention measuring the concentrations of chemicals in the air. Geographic information systems and satellite information are also used to assess the degree of exposure to air pollution in addition, to air pollutant ground monitors.[47] On the other hand, technologies for monitoring physical activity by using tools attached to the body, and those for continuously monitoring physiological information, such as blood pressure and pulse, are already being used.

In the future, technologies for monitoring environmental exposure and for physiological and pathological evaluation of body systems will be further developed, thereby facilitating the use of technologies for comprehensive monitoring based on individual persons rather than on the evaluation of individual factors. As smartphones are evolving from simple mobile phones to more advanced devices, including providing access to the Internet and functioning as a game console, a music app, a camera, and other devices, these technologies shall be further developed to enable the monitoring of physical activity, physiological responses, dietary intake, and exposure to pollutants via body attachment tools or via hyperconnected networks in our daily life. Moreover, this information can be useful for the health management through integration and interpretation of the data in terms of the harmony and balance between the human body and the environment, rather than just gathering individual pieces of information. In addition, the information obtained as such will be linked to the hospital's health management system, which will again enable the more effective prevention and treatment of diseases.

The flow of time affects the development of a disease

We are exposed to a number of different disease risk factors throughout our lifecycle, from the time of our conception to the time that we become infants, children, adolescents, young adults, middle-aged people, and elderly people. It is also important to assess not only the time of exposure to these risk factors but also their relevance to the health effects that occur after considerable time has elapsed. In fact, the reason for assessing the association between disease risk factors and their effects on health is to analyze and understand networks of highly complex entangled systems to prevent and treat disease. The problem, however, is that the current disease risk factors are more likely not related to the current disease but to future health effects. For example, exposure in the mother's womb and during childhood can develop a health consequence after middle age. As shown in the Dutch Famine Study, a fetus that is not properly fed in the womb develops various chronic diseases, such as obesity, hypertension, diabetes mellitus, and cancer, when the child grows and becomes middle-aged.[48] Therefore, it is important to take measures to prevent diseases that are likely to occur in the future by analyzing various disease risk factors before the disease occurs.

When there is a considerable time gap, however, between the exposure to a certain factor and the eventual occurrence of the disease, it is more plausible to assume that gradual changes in the human body have been accumulated from a young age before manifesting themselves as a disease in the middle age or later, when the ability to recover to the normal state falls apart, rather than to believe that exposure in childhood suddenly develops into a disease in later life. Therefore, there is a need for an overall assessment of the exposures that occur during the fetal and infantile periods, a very sensitive period to disease risk factors, as well as for a constant monitoring of the changes in the structure and functions of the human body over the lifetime. This can help prevent disease later by identifying and correcting the problems in the body as early as possible.

Social relationships also begin with the relationship between the pregnant woman and the fetus and evolve into the relationships among the family members, the relationship between the teacher and the students, and the relationship between employer and employees. It also lies in the complexly entangled network of relationships, such as in the social order and class or hierarchy. Category of classifications such as religion, ethnicity, and race also represent very important social relations. Such a network should in fact be considered a consolidated system, one where the relationships are linked with one another so closely that it is very difficult to grasp any of them independently.

Moreover, the network is not fixed at any moment but undergoes temporal changes depending on the time along the lifecycle. Life itself undergoes a change from birth to death through the process of growth, development, and degeneration to death. As mentioned earlier, exposure in early life may appear as a disease in later life, and temporal change is not merely a change but one implying the causality with the change of time. A range of external system factors affect the human body in a multidimensional manner, and the human body reacts to the external system factors through the changes in the genes, epigenetic programs, protein expression, and immune or inflammatory response in the course of time. If such reactions stray out of the normal range of physiological changes, they manifest themselves as a disease at that time.

On the other hand, the change of overall society is so great that one can see that the exposure to disease risk factors, whatever they are, has undergone considerable changes over time as well. Compared with a decade ago, people's dietary habits have changed, and the concentrations of environmentally harmful substances such as heavy metals have also changed. Housing, transportation, and environmental sanitation have changed as well, along with the possible changes in the microbial flora in the intestines and in the living environment. This time-varying exposure pattern, coupled with the individual's lifelong physiological changes, further complicates the relationship between the exposure to risk factors and disease development. This temporal change must be considered as another complex system. Therefore, it is necessary to properly evaluate the networks between the systems that change over time to enable accurate disease prevention and treatment.

Systems medicine approach is needed

As such, a new medical model for diagnosing and treating diseases based on intersystem networks can be called the "systems medicine model." The systems medicine model basically starts with the concept of the harmony and balance of the whole system. Therefore, it is an organically integrated approach that goes beyond the mechanistic and analytical rationalism. The systems medicine model is based on the idea that disease is caused not by a simple causal factor but by the disrupted balance of the multidimensional systems network inside and outside the human body. In other words, the systems medicine model assumes that a disease breaks out when the balance is disrupted across varying levels, from the balance between the human and external disease risk factors to the balance between humans and microorganisms such as bacteria that are present in the human body. Further, it also extends to the balance between the cells constituting the tissues in various organs of the human body as well as between intracellular organelles and the mitochondria, thereby hindering the normal functioning of the human body. In addition, as the social environment created by humans is rapidly changing as well, it is understood that when one fails to adapt well to the changing social environment, the balance of social relations is also broken, triggering development of mental diseases such as depression as well as various diseases affected by stress.

This view of diseases based on the systems medicine is inevitably different from the biomedical view of disease in approaching disease prevention and treatment. While the discovery of the specific cause of a disease and of its prevention or treatment is central to maintaining and restoring the health according to the biomedical view of disease, the systems medicine approach's view of disease assumes that maintaining and restoring the balance inside the cells, among the cells, between humans and microorganisms, between humans and the environment, and in human social relationships is key to preventing and treating a disease. These two different views of disease, however, are not necessarily irreconcilable concepts. There are many cases where finding and eliminating the specific cause of a disease is crucial to maintaining or restoring the balance.

As an example, let's look at the blood glucose test that is taken when treating diabetes mellitus. According to the biomedical view of disease, an appropriate medical measure is to treat high blood glucose level with hypoglycemic agents or insulin. According to the systems medicine's view of illnesses, however, diabetes mellitus is the result of the disrupted balance between normal energy production and consumption owing to failed dietary management, lack of physical activity, stress, cigarette smoking, and exposure to chemicals. Furthermore, it is also closely related with living conditions such as increasing food consumption, living infrastructure such as cars and elevators, and air pollution. Therefore, the right medical measures according to the systems medicine approach are to restore the balance between energy production and consumption through the comprehensive improvement of the lifestyles of individuals, the improvement of the environment in the residential communities, and adequate medical treatment to lower the blood glucose level. Thus, medical measures also include medication, such as hypoglycemic agents, just like the clinical practice based on the biomedical model. Taking hypoglycemic agents, however, is only one way of restoring the balance of energy production and consumption in the treatment of diabetes mellitus according to the systems medicine's view of disease.

In the systems medicine approach, even community policies to promote health should not be applied invariably to all situations or to everyone but should be applied by carefully considering each situation or by analyzing the relationship of the internal systems of each person. Let's take, for example, the policy of walking or riding a bicycle when going to school to increase physical activity and reduce obesity. This increase in physical activity is not always good for the health of the body, however, because it can adversely affect the health, for instance, when the air pollution is severe. Therefore, for such policy to benefit the health, the air pollution level should be constantly monitored, and each individual should be given the right information regarding the air pollution level. On the other hand, the effect of exposure to physical activity or air pollutants on health can also vary among individuals, even in the same level of exposure. Therefore, it is necessary to take into account the roles of genomes, epigenomes, proteins, and metabolites as well as current and past illnesses, which cause various reactions when exposed to hazardous substances such as air pollutants (e.g., irritation, inflammation, and immune and metabolism reactions), before taking the necessary actions. Foods or nutrients as supplements to strengthen body defense, or in some severe cases, medicines such as anti-inflammatory agents or immune modulators, may be used. The use of such foods or medicines is not limited to specific health effects but is targeted at multiple mechanisms of the complex internal systems to maintain the harmony and balance of the overall system.

When you are confronted with a health problem, you may also be able to solve the problem by targeting the microorganisms that are in a symbiotic relationship with the human body or by working with microorganisms. Microorganisms are independent entities, but because they are in a symbiotic relationship with the human body, they can be seen to be interconnected with the various systems of the human body, which results in a single integrated system. Most microorganisms are residing in the small and large intestines and contribute to the production of energy by breaking down the food that we ingest. They are not only involved in obtaining energy, however, but are also directly involved in the immune system. It was recently discovered that they are associated with chronic diseases or immune disturbances like obesity, diabetes mellitus, and Crohn' disease. In fact, there are hundreds types of bacteria only in the intestines, and their distribution and interaction with the human systems form another complicated network. Therefore, although microorganisms are yet to be more evaluated for their roles, they are also an important factor to be considered for the systems medicine approach.

From standardized treatment to customized treatment

In this way, systems medicine is a medical approach aimed at comprehensively understanding and managing the complexity of the relationships between the molecules, cells, and organs that constitute the human body, as well as between humans, microorganisms, and even ecosystems. Each element of a system involved in the relationship plays an independent role but also serves as a link between systems.[49]

The *APOE* gene, for instance, is involved in cholesterol metabolism, providing cholesterol from the food to the body for its use through blood flow. In modern humans who consume excessive calories, the *APOE* gene, which had been adapted to the food intake during the hunter-gatherer period, plays a role to elevate the risk of atherosclerosis. When there is a mutation in the *APOE* gene, atherosclerosis can be developed more easily by the mutation acting in ways to further increase the cholesterol level in the blood compared with the case where there is no such mutation. Atherosclerosis per se, combined with factors that cause inflammation in the blood, such as cigarette smoking and air pollution, is likely to cause thrombosis, and if thrombosis occurs, it can block the blood vessels that supply blood to the heart, causing heart diseases such as myocardial infarction. In this case, the *APOE* gene or the mutations in the gene interact with factors such as food, smoking, and air pollution. Here, food is related to regional and cultural factors, smoking to lifestyle, and air pollution to industrialization. Ultimately, heart disease is caused when factors such as genes, food, smoking, and air pollution, and their links between them, fail to form harmony and balance with the systems that make up the human body.

Even though we explain the causal relationship of the *APOE* gene and heart disease with this level of relevance, however, it will also be an oversimplification of the original complex relationship. In reality, food intake affects the action of the *APOE* gene via the epigenetic regulatory mechanisms, which affect the production of proteins via RNA, then dozens or hundreds of proteins are mobilized, and various enzymes or inflammation mediators are also involved in various human reactions. Meanwhile, as food is digested and metabolized, a large number of diverse metabolites are formed, some producing toxic or inflammatory effects and others providing defense against them. In addition, to food intake, cigarette smoke and air pollution also contain more than a 100 different chemical substances and trigger various reactions in the human body. Accordingly, it can be said that the relationship between the *APOE* gene and heart disease needs to be understood in such a very complex relationship.

Therefore, first, to understand such a complicated relationship, it is necessary to analyze different human systems, such as the genome, epigenome, proteome, and metabolome. A network can be understood as a functional and hierarchical relationship within and between systems. As mentioned earlier, as these networks are complex, dynamic, and changing over time, there is clearly a limit to approaching them with an a priori hypothesis. Even if a hypothesis is well constructed, it is desirable that a series of new relationships emerging from the data continue to be added to the hypothesis so that the hypothesis can be improved over time, therefore enabling the network analyzed correctly and evaluated thoroughly. In other words, only through a combination of the "hypothesis-driven approach" and the "data-driven approach" can the network of complex systems be accurately analyzed to clearly see the causality of factors that affect disease development and progression.

Until now, modern medicine has been narrowing the scope of inquiry from the organ to the tissues and cells, from the cells to the organelles in the cells, and further down to the level of molecules, such as the DNA or RNA in the nucleus. In the systems medicine approach, however, these subtle factors are not regarded as independent but are thought to be connected one another and present in various forms in each individual or patient in the context of hierarchical or interrelated network. Therefore, to perform treatments according to the systems medicine approach, the conventional diagnosis method that differentiates diseases from one another based only on clinical phenomena, such as symptoms or pathological findings, needs to be revised. The addition of microscopic or molecular profiles of the genes or proteins, and other molecular biomarkers, will lead to a much more detailed and precise diagnosis of the disease. If a precise diagnosis is made, it will be an important basis for the advancement from the current standardized treatment method, in which all patients receive the same treatment if they belong to the same disease group, to the customized treatment tailored for individual patients.

A step closer to the conquest of disease

As the DNA code of all the genes can be easily analyzed and as the technology for analyzing the epigenome, the proteome, and the metabolome has been sufficiently developed, it is now possible to evaluate not only the sequence and the variation of the genes but also their functions in linkage with disease occurrence or progression. The mathematical and statistical models of how these complexly entangled changes are controlled with one another and how they function together, as well as related technologies capable of closely monitoring the biological changes in the body, have been made possible by the development of computer and life science technologies. Of course, we have yet to fully understand the physiological mechanisms behind the biological process of birth, growth and development, and aging, but we will be able to gain a deeper understanding of the lifelong changing processes, as well as of disease, through the systems medicine approach soon.

The systems outside the body in the perspective of systems medicine are all potential disease risk factors that can affect one's health from conception to death. Therefore, not only the degree of exposure to individual factors but also the extent to which the exposure changes over time should be considered. Individual factors may refer to lifestyle, environmental pollution, microorganisms, occupation, social relations, and other elements. Each of these factors, however, is actually a component of very complicated system. These exposure factors, independently or in conjunction with one another, activate or stimulate the human reaction systems, such as the immune, detoxification, and inflammation systems of the human body. These reaction systems will be described in greater detail in Chapter 3, but they are not at all simple; on the contrary, they are also highly complicated because they have been developed over a long period of time through the natural selection process to defend against various threats of life.

The history of humankind, however, shows that people behaved well in the age of hunter-gatherers by recognizing the patterns in their complexly entangled environment. For example, hunting an animal requires not only knowledge of the location, the characteristics, and the condition of the prey but also knowledge of the threats surrounding it as well as consideration of the hunter's own strength and availability of food, water, and weapons. To understand and address these complexities, the human brain has grown larger over time and has attained its current size and structure. Complexity, therefore, is not a new and unfamiliar issue but has been a very familiar subject for humans. Equipped with the sufficient capability to process information by remarkable technology development, the systems medicine approach will become similar to our brain's own approach. If humanity advance information processing ability more in the near future and particularly get help from artificial intelligence, it will be possible to make more accurate diagnoses and to better manage people's health by using the advanced techniques. In other words, when various pieces of information are available in real time, the computer program equipped with such capacity can determine the factors affecting your health in real time as well. We can perform precise health management based on this information, suggesting that we will have a state-of-the-art capability and a technical system that can provide the best health care in real time, as necessary, propelling us a step closer to the conquest of disease soon.

Chapter 3

Five Strategies of Our Body to Fight Disease

3.1 Humanity to Live With Microorganisms in a Symbiotic Relationship

The process of mixing different individual organisms, such as symbiosis, gene mixing, or cell association, was a stepping stone to the evolutionary leap forward. For instance, the mitochondria were created as a result of combination of certain bacteria with archaea. This way, a system with proper differentiation of the nuclei and mitochondrial functions was established within the cell, which again formed the basis of a complex system of humans. In addition, the symbiotic microorganisms have become indispensable for human survival as they perform crucial tasks, such as digesting food or blocking pathogens. Therefore, it is important to understand the diseases that are caused or aggravated when the harmonious or balanced symbiotic relationship between the human body and the microorganisms is breached and to treat diseases in ways that would sustain such symbiotic relationship.

Leap of life founded on coexistence with other species

All multicellular organisms, such as fungi, plants, and animals, evolved from single-celled organisms like bacteria or archaea. A deeper examination of these evolutionary changes shows how independent cells exchange information and combine with one another to become multicellular organisms. Of course, unicellular organisms, such as bacteria, would have initially evolved from some other organisms. It can be inferred that equipped with nucleic acid called "DNA," which is composed of codelike base sequences such as A, T, G, and C, the first creature that existed had the ability to produce the next generation. It is difficult, however, to know how the first creature came to be changed before evolving into a very complex system of mammals, and eventually into humans. What is clear is that the gene alone did not promote the evolutionary changes. The most likely scenario is that genetic change took place with the force of natural selection, in which living organisms strive to survive by adapting to the given environment. In other words, it can be concluded that the environment dictated the direction of the evolutionary changes that organisms underwent.

The Changing Era of Diseases. https://doi.org/10.1016/B978-0-12-816439-6.00003-X

It can be said that the interactions between the genes and the environment have led to the changes collectively called "evolution" through natural selection. It is difficult to understand all the phenomena involved in evolution, however, with this explanation alone. In fact, the process of selecting a gene that fits the environment better through natural selection may be tantamount to determine life and death at the individual level, but when seen at the group level, it can be understood as a process of gradual changes rather than an abrupt change. The change that occurs in each individual usually contributes to gradual changes that occur in the characteristics of the species over time rather than is a single event that changes the characteristics of the species at once. As the birth of a new species, however, signifies the emergence of a species with different characteristics from the previous ones, the process of gradual changes alone, such as natural selection, may not be able to sufficiently explain the emergence of a new species.

In the process of evolution, there are certain changes that occur that would mark a qualitative leap into a new species, such as changes from prokaryotic to eukaryotic cells, as well as changes that further increase the adaptability of a given species. Therefore, another powerful propulsion system for evolution is required to explain this phenomenon: "symbiosis" of species or a mixture of their genes. These associations are usually made between close species, but they are sometimes made through the processes of eating and being eaten by distant species, such as in the formation of mitochondria and chloroplasts of eukaryotic cells. The birth of a new species may be the accumulated result of the repeated natural selections of one species, but it can be said that the symbiosis facilitated by the cooperation between or the combination of heterogeneous species also plays a significant role in it. Therefore, if evolution is regarded simply as a natural selection process through individual competition, there will be a limit to understanding various biological phenomena because there have been occasions when the cooperation among individual organisms, such as in symbiosis, in addition, to individual competition, have created better species.

The mixing of genes also plays a very important role in evolution. Humans are born essentially through the process of combining the individual cells of the sperm and the egg into one, suggesting that from the beginning humans have been born through the process of mixing genes. The human may be one of the most advanced multicellular organisms, but it actually begins with a single cell created through the association of the egg with the sperm. After that, the single organisms are divided and differentiated into multicelled organisms, which, again, eventually evolve into a human being. This process can be viewed as one of the reuse of the cell union programs that occurred when the single-cell life forms on Earth changed into multicellular life forms.

In fact, if we look at the evolutionary process from a single-cell organism to a multicellular individual organism, we will see that even for different species, they usually go through a similar process. If the process of changing from a single-cell organism to a multicellular organism is similar in all cases, however, there should be a common reason for the development of multicellular organisms in the world where only single-cell organisms were prospering in the past. Perhaps when a single cell was initially divided, there could have been an occasion that the cell did not divide into two but had become two single cells joined together instead. The cells enlarged as such would have been beneficial in competing with other single cells. Perhaps their larger size reduced the likelihood that they would be eaten by other predators. This might have led to size competition, setting the evolutionary wheel in motion toward multicellular organisms.

When a single cell becomes a multicellular cell as such, however, the weight of the cell increases, and the efficiency of cell movement decreases. In addition, as the cell's surface area is smaller compared with the cell's weight, the cell may face limitations when ingesting nutrients or discarding waste using its surface wall. Eventually, various changes might have occurred to overcome this limitation, which might have caused cell differentiation. When equipped with motion device, such as the flagella (that can be considered the feet of a cell), the cell would have been a favorable condition for natural selection because it would have the ability to move easily. The emergence of digestive organs that ingest and excrete nutrients has also become a very important change in terms of energy use. On the other hand, being a multicellular organism means that complex systems like communication and control systems between cells have begun to be formed in oneself, and that one has come to develop complex functions, including neural networks that can efficiently transmit information in the body. Over time, multicellular organisms have become more sophisticated, adding new features like metabolism and nerve reflexes.

Mitochondria were born thanks to the symbiosis with different bacteria

The process of mixing different individual organisms, such as symbiosis, gene mixing, or cell association, becomes a stepping stone to the subsequent evolutionary leap forward. In fact, bacteria exchange genes frequently between themselves, which again causes genes to move around among bacteria. In this case, the bacterium may have not only the genes that it received from the bacterium from which it was divided but also another gene that it received from another bacterium. Therefore, over time the genes of a bacterium may become different from those of its direct ancestor. This signifies the birth of a chimeric bacterium, whose genes are very different from those of its ancestral bacterium due to gene blending.

The chimeric mixture of a prokaryotic bacterium faces a decisive moment when a eukaryotic cell was born! The events in which the mitochondria become the "energy powerhouses" of the cells represent this mixing process. The mitochondria were created as a certain bacteria combined with other microorganisms. What was surprising in the event was not the gene mixture or symbiosis between bacteria of the same kind but between archaea and bacteria—that is, the combination of different species of unicellular organism. As such, the birth of eukaryotic cells, considered the most important event in the history of evolution to a higher organism, is the result of the incorporation of bacteria and archaea into a single cell. Eukaryotic cells are higher functional cells that are equipped with both mitochondria and nuclei. In other words, eukaryotic cells have come to possess mitochondria in the cytoplasm, which produce energy while acting like independent organisms inside the cells, along with the nucleus, which acts as a command center controlling the functions of the cells, thereby shifting the evolutionary wheel of life to a high gear.

Although mitochondria have an independent bacterial form, a considerable portion of the mitochondrial gene is transferred to a nuclear gene located in the nucleus of a eukaryotic cell, leaving only a few genes in the mitochondria. This represents a mixture of genes from the standpoint of the nuclear genes and a loss of genes from the standpoint of the mitochondrial genes. If most of the mitochondrial genes, however, migrate to the nuclear genes, why is it that not all of the mitochondrial genes migrate to the nuclear gene and some genes still remain in the mitochondria? Nuclear genes can be safely stored in the nucleus as well as in the proteins called "histones." As such, mitochondrial genes may have a reason if they remain in a dangerous place that produces energy constantly, just like a power plant.

There are hundreds of mitochondria in the cell, and mitochondria play a very important role in converting the nutrients entering the cells (e.g., sugar) to energy. If the energy production is to be controlled according to the supply and demand of energy, however, it may be better for each mitochondrion to independently fine-tune the energy production than for all the mitochondrial functions to be controlled indifferently by the nuclear genes, just as individual heating systems working independently are more efficient than centralized heating in the building. This way, a system with proper differentiation of the nuclei and mitochondrial functions was established within the cell through events of heterogeneous species combination and gene mixing, which again formed the basis of a complex system that the advanced organisms like humans have come to possess.

Humans and microorganisms: indispensable companions for each other

The genes that have once adapted to the existing environment may have difficulty adapting to the changing environment when environmental changes occur in a short period of time, and such genes' maladaptation may lead to disease occurrence, particularly chronic diseases. There is a limit, however, to explaining diseases only by the environmental maladaptation of the human genes. There is no doubt that environmental change is the decisive causal factor of chronic diseases, but it is not only the human nucleotides and mitochondrial genes that fail to adapt; other living organisms in a symbiotic relationship with humans also cannot adapt well to the changing environment.

Therefore, it is necessary to consider the genes of whole organisms, including those in a symbiotic relationship, when thinking about the maladaptation of genes. In other words, the genes in the nucleus must be evaluated together with the genes of the mitochondria in the cytoplasm or the genes of the bacteria living in the intestines to determine in what way each gene is maladaptive to the environment. Perhaps it would be more appropriate to look at humans as symbiotic complexes rather than as single independent living organisms and to comprehensively evaluate their genes of a symbiotic system as a whole.

This symbiotic relationship can be better observed by looking at individual humans. As we have 10 times as many microorganisms in our bodies as human cells, we can say that 90% of the cells in our bodies are microorganisms like bacteria.[46] These microorganisms have become indispensable for human survival as they perform crucial tasks, such as digesting food or blocking pathogens. As microorganisms also obtain the basic necessities for survival from humans, there exists a truly symbiotic relationship between microorganisms and humans that makes them indispensable to each other. In fact, not only humans but also the vast majority of biological species depend on such symbiotic relationships.

How then can we define human "individuality" or an individual human being? Genetic analysis shows at least a 1000 different species of bacteria living in the human colon alone. In addition, there are many different bacteria living in the human skin, mouth, esophagus, and genitalia.[50] If we look at it this way, we can say that a human is a creature in which the human body parts equipped with human genes and the many other different microorganisms are living together while forming a symbiotic system. What if there is no such symbiotic system? Rats born and raised in the absence of germs were found to have a poor immune system and poor gastrointestinal development.[51] In short, it is difficult to sustain life without the symbiotic system.

In fact, symbiosis is not an exceptional phenomenon but simply another way for achieving evolution, thereby posing some fundamental questions regarding the very concept of an individual entity. Until now, individuals have been recognized as independent organisms, and there has been no problem with seeing them as standalone units in academic disciplines from biology to philosophy. Many scientific and philosophical themes have so far focused on individuals, and even to this day, we see objects by focusing on individuals. Now, however, it is reasonable to regard human beings as living organisms in the context of "symbiotic complexes" or "extended individuals," away from the traditional concept of seeing them as individuals. Therefore, symbiotic microorganisms should be evaluated together with the human body when evaluating the factors affecting disease development.

We get sick when our partnership with microorganisms breaks down

Ultimately, disease is a phenomenon derived from the process of evolution over a long period of time. Therefore, even if we understand disease as a pathological phenomenon, we need to look at it from a historical point of view. Only then can we properly understand the impact of civilization on the outbreak of disease today and take a better position for managing it. The domestication of animals, which began with the emergence of civilization, created an environment where wildlife animals and human beings lived together, deepening the interdependence between them. For example, chickens have never been as numerous as they are now. As chickens were being raised by humans, the poultry population greatly increased in number, and their survival became absolutely dependent on humans. In addition, humans began to eat chicken and eggs to ingest protein, fat, and other nutrients, and it is now hard to imagine a dietary life without chicken and eggs. The domestication of animals may have started in such an attempt to secure food stability, but the domestication of the chicken became also a form of symbiosis of humans and chickens. This symbiosis, however, also brings about exchanges of microorganisms between humans and chickens.

The infectious disease caused by the avian virus can be regarded as a phenomenon occurring out of such symbiotic relations between humans and poultry. The mass rearing of poultry including chickens and ducks made it easier for the birds to get infected just like urbanization emerged as a key cause of infectious diseases after the Industrial Revolution because of the clustering of people. Just as pest germs have caused widespread epidemics among humans after changing their hosts from wild rats to house rats, then to humans, if the pathogens change their hosts from wild birds to mass-bred chickens or ducks, they can cause an epidemic of avian influenza among the poultry. Moreover, if pathogens change their hosts from birds to humans as another host, avian viruses will be able to cause epidemics in humans.

Therefore, not only is the domestication of animals a process of symbiosis between humans and animals, but it can also be referred to as a "search process for a possible symbiosis of humans and microbes that took animals as hosts." Of course, there can be unpredictable results at the time of the search, and during the period, new viral or bacterial diseases can lead to unmanageable outcomes. Indeed, the search for a possible symbiosis is not unusual in earth's long history, and until the complete symbiotic relationship is established, diseases can develop anytime. For instance, avian viruses cause diseases in birds because the complete symbiotic relationship between the avian viruses and birds has not yet been established. If a symbiotic relationship is established, there will be no reason for the avian virus to kill birds. Furthermore, it is possible for the avian virus to mutate into a form that can infect humans and be propagated among humans, but then again, this can also be regarded a search process for a symbiotic relationship between humans and birds or the avian virus.

In fact, many microorganisms in the human body have formed a symbiotic relationship with humans. When humans are first infected by such microorganisms, the latter become quite virulent, often causing disease and death. As the microorganisms with less virulence flourish, however, they cause increasingly fewer diseases and eventually reach a stage where they would benefit from coexisting with the human cells. For instance, in the case of infection by bacteria, various gene mutations are supposed to develop among the bacteria that cause infection. If bacteria with a weak pathogenic capacity are more viable than those with a strong pathogenic capacity, the former will be selected in the end. The greater the pathogenicity of a bacterium, the more likely it is for such bacterium to endanger the life of the host or to cause its death. Since bacteria cannot survive when their hosts die, this will be also a threat to the survival of the bacteria themselves. As such, bacteria can change their pathogenicity through the mutation of their own genes while they can engage in gene exchanges with other bacteria.

The formation of a symbiotic relationship with humans is evident in the case of bacteria, but it is also likely in the case of viruses. Therefore, HIV, the Ebola virus, the avian virus, and other viruses that threaten humankind may no longer cause disease to humans at some time in the future because they may eventually become symbiotic viruses. Of course, symbiotic relationships are not formed in all cases. If humans develop a vaccine against a virus or a bacterial pathogen, or if immunity naturally forms, there will no longer be cases that the pathogens enter the human body and cause disease even without a symbiotic relationship between such pathogen and humans.

On the other hand, when an organism is infected by a virus, the virus gets an opportunity to acquire some of the genes in the cells of the organism, suggesting that the virus can become a mediator of interspecies gene exchange. For example, if a virus that infects an animal has an opportunity to insert a piece of that animal's gene in its own gene, and the virus infects humans later and delivers its gene fragment to a human gene, the gene has been transmitted between different species. As such, genetic interactions through viruses can occur between organisms that are systemically separated from each other. In fact, there are no species or individuals that consist entirely of different genes from other species. All organisms basically come from such gene blending. Symbiosis or genetic mixing, therefore, is not an exceptional phenomenon of organism creation and evolution, but it may be more appropriate to view it as a basic principle. Therefore, when trying to understand the state or phenomena of an organism, such as health or disease, symbiosis or gene mixing must be considered as well.

Symbiosis with microorganisms, the key defense strategy of the human body to prevent disease occurrence

It can be said that the relationship between pathogens and their hosts is meant to maintain a balance. In other words, if pathogens multiply at rapid speeds in a human body and cause severe illness and increase mortality, the pathogens can initially spread well, but if the infected person dies or cannot move, the pathogen may lose an opportunity to spread to other humans anymore. On the other hand, if the pathogens breed slowly in a human body and do not cause a severe illness owing to their relatively weak toxicity, they may spread slowly in the beginning, but all told, they will have a greater chance to propagate among humans. Although pathogens take one of these two strategies, they usually switch from the first strategy to the second one before ultimately choosing an entirely new strategy. In other words, pathogens escalate their relationship to a symbiotic one in which they and their host gain mutual benefits from each other. The relationship between the predator and the prey may look like where one party is being sacrificed unilaterally, but if the prey decreases, the predator decreases accordingly, and as the predator decreases, the prey increases again, which results in increase of the predator again.[52] As such, the predator and the prey form a relationship in which they can coexist with each other. Although each individual may practice these principles without necessarily understanding them, the principle of symbiosis, which ensures harmony between two parties, with the benefit not becoming lopsided towards any specific party, is being realized in the ecosystem.

Every living organism is born with the purpose of transferring its genes to its offspring. Thus, an individual is a unit that delivers its genes to its offspring. When an individual, however, refers to an organism that is in a symbiotic relationship with microorganisms, its genes cannot be defined that simply because the organism requires genes of microorganisms as well as the genes of itself. It can be considered, therefore, that the host selected by the force of natural selection is selected not because of the excellence of its genes but because of the superiority of all the genes, as the host and microorganism are seen as one symbiotic complex. This is a new concept that differs from the previous assumption that the genome of the host is transferred to the next generation and that, in the process, adaptation to a new environment is enabled by the pressure of natural selection working on the host gene. This new concept of natural selection working on symbiotic relationship can have a significant impact on the disease treatment strategies as well as on the understanding of diseases.

Microorganisms like bacteria, viruses, and fungi, which are present mainly in the human epithelial cells such as the digestive, respiratory, skin, and reproductive systems, have recently been reported to be closely related to disease occurrence. The composition of microorganisms is known to vary from one person to another, and the composition variations are reported to be related to various diseases, such as autism, ankylosing spondylitis, enteritis, and obesity. For instance, Larsen et al. have found that diabetes mellitus is associated with the decreased diversity of intestinal bacteria. In addition, an analysis of the nucleotide sequence of about 700,000 intestinal bacteria revealed that diabetes mellitus and obesity are closely related to the ratio of bacteria in the Firmicutes and Bacteroidetes families.[53]

The reason that the change in the bacteria distribution can increase the risk of disease is that the change in the distribution of the bacteria into specific compositions that considerably deviate from the symbiotic relationship may affect the immunity or metabolism function of humans. Symbiosis is an important defense mechanism that is created in close cooperation with other living organisms. Therefore, if the partnership is broken or is not running smoothly, the defense capability will deteriorate as well, undermining the defense against external factors. Many of the diseases that occur in the human body can be seen as a result of this broken symbiotic relationship.

Microorganisms, therefore, are not only involved in the production of energy and nutrients in the body but are also constituting an important system making up the human defense system, such as immunity. It is therefore important to understand how these microorganisms are organically linked and how they interact with external factors like a person's dietary habits and chemicals, as well as with the other defense systems within the human body, such as immunity and metabolism. In addition, it is important to understand the diseases that are caused or aggravated when the harmonious or balanced symbiotic relationship between the human body and the microorganisms inside it is breached, and to treat diseases in ways that would sustain such harmonious or balanced symbiotic relationship. In other words, it is necessary for people to improve their food intake so that the normal microorganisms in the intestines can be strengthened or maintained in an optimal state, or to administer microorganism drugs when necessary, so that the body's symbiosis with the microorganism could prevent or cure the disease. To successfully manage diseases, it is important to understand the impact that such symbiotic relationships have on the health and on disease and to use such knowledge wisely.

3.2 Strengthening the Body's Defense Against Toxins

The human poison metabolism system is quite flexible and acts in a well-coordinated manner. Even this excellent poison metabolism system, however, cannot prevent all the attacks of toxins. The detoxification mechanism is designed to decompose the toxins and to make them soluble in water, before discharging them through the kidney into urine. Oxygen is used to convert lipophilic toxins to water-soluble ones, producing reactive oxygen species as byproducts in its wake. Therefore, the greater the exposure to toxic chemicals, the greater is the amount of reactive oxygen species produced, causing oxidative stress, which is the central factor that causes or exacerbates various chronic and late chronic diseases. Efforts are needed to maintain the balance between the production of reactive oxygen species and the activity of antioxidant systems.

The flexible poison metabolism system in our body

Biological creatures coexist and rely on one another for survival, but they also fiercely compete with their perceived enemies for survival. About 500 million years ago, when living creatures migrated from the sea to the land to settle down, plants chose to settle on the land without moving. This is because plants did not have to move to get food because they could get the energy from the sun due to the chloroplasts that entered their cells and then established a symbiotic relationship with them. Plants, however, lacked mobility and needed a way to stop the attack of animals or insects. Faced with such problem, plants developed defense mechanisms to prevent extinction by being eaten by enemies, one of which is to produce poisonous or irritating chemicals. Examples of these plants include mushrooms containing poison that can kill the mushroom hunters when they are eaten and plants that cause allergic reactions like urticaria when they come into contact with the human body. On the other hand, through the process of natural selection, animals have also developed a poison metabolism system to mitigate the toxicity of plant-generated chemicals as many animals cannot survive without eating plants.

The term "detoxification" has been widely used of late in the sense of eliminating harmful effects. It is often used to signify the act of eliminating a harmful effect in our bodies, such as drinking detoxifying juices and bowel cleansing. The word "detoxification," however, originally refers to the poison metabolism system that converts harmful substances into harmless ones. It usually means a process in which various enzymes in the liver are mobilized to neutralize toxins and then to dissolve them in the blood before releasing them from the body through the kidneys. This detoxification process is the most important mechanism of the body for removing unfamiliar or hazardous chemicals when these come inside the body from outside.

We are exposed to various chemicals through our ingestion of foods and drinks or through our contact with the environment in our daily lives. Many of these chemicals, however, are different from those that our hunter-gatherer ancestors were exposed to in the past. In particulary, we are not only exposed to entirely new toxins that had been produced since the Industrial Revolution that are quite different from the chemicals to which our hunter-gatherer ancestors had been exposed; we are also exposed to much higher amounts of such toxins. The vast majority of poison metabolism systems that people now possess, however, were formed a long time ago before the Industrial Revolution. In other words, most of such poison metabolism systems had been formed during the age of hunter-gatherers, and only a little had been newly equipped through genetic mutations since the beginning of civilization, such as the lactose metabolism capacity related to milk intake.[54] Therefore, the chemical substances that people are newly exposed to have to be metabolized mostly through the existing poison metabolism systems, which had already been established by the age of hunter-gatherers.

What is surprising, however, is that the human poison metabolism system has some capability to detoxify even a new chemical entering the body. This is because the poison metabolism system is quite flexible as it is composed of complex programs in which several different proteins act in a well-coordinated manner in a series of stages. Even this complex and excellent poison metabolism system, however, cannot prevent all the immense amount of attacks of toxins. Some toxins may express their toxicity immediately, even before going through the poison metabolism system, and some other substances' toxicity may escalate while they are going through the poison metabolism system.

On the one hand, the poison metabolism system also plays an important role in releasing the residual hormones, neurotransmitters, vitamins, and inflammatory substances out of the body after being utilized for various purposes.

If these substances remain in the body after the use, our bodies may end up losing their control function over such substances. Therefore, the poison metabolism system does not simply remove the toxins from the body but secures the balance of the substances required by the human body. It is one of the central defensive strategies crucial to maintaining health by reducing the toxicity of exogenous or internally generated substances. If the poison metabolism system does not play its given role, chronic diseases like liver disease, neurodegenerative diseases, and diabetes mellitus can be caused or exacerbated. Therefore, it is very important to take full advantage of these defense strategies and to provide the necessary support to the system to work properly for the prevention and treatment of diseases.

The working capability of the poison metabolism system largely depends on a person's lifestyle, the degree of exposure to the environment, and his or her genetic characteristics. For example, if your liver is not functioning well due to too much alcohol intake, your poison metabolism system's performance in terms of removing the toxic chemicals from your body will be lowered, causing you to absorb a greater toxic impact from chemicals than would people with a normal liver function. Furthermore, as the genes involved in the ability to metabolize alcohol or toxic chemicals show polymorphisms frequently among the population, a person's genetic characteristics also play an important role in this case. Therefore, the assessment of toxicity can be done properly only by considering the person's genetic, metabolic, and clinical conditions and assessing the person's lifestyle as well as all the substances that the person is exposed to in his or her living environment.

The proteins in the body work together to remove toxins

Plants have developed defense mechanisms, such as poisoning the animals that are trying to eat them or causing such attackers confusion by wreaking havoc with their nerves. On the other hand, animals not only have to overcome such toxicity of plants but also use them as energy sources or nutrients by eating them. Animals have been equipped the capacity by developing enzymes that can remove the toxicity of the chemicals emitted by plants for their own protection. In particular, the "cytochrome P450 enzyme system," which forms the basis of the chemical metabolism system, is a complex and elaborate poison metabolism system, which has been equipped as a result of a long period of evolutionary time to detoxify numerous plant poisons.[55] In fact, the cytochrome enzyme system consists of a protein family with the highest variety among the protein families. Therefore, this enzyme system can deal with some of the new chemical substances that appeared after the Industrial Revolution. The problem is that even such an elaborate and flexible system is not capable of handling all chemicals perfectly.

The metabolic process of the cytochrome enzyme system begins by attaching oxygen to the chemicals that were introduced to the human body from outside. This is to make the chemicals easily soluble in water so that they can be discharged from our bodies. However, the process leading up to the release of the chemicals is not necessarily safe or simple. Let's look at the toxic chemical called "benzene." When benzene enters our body, the cytochrome enzyme system attaches oxygen to the benzene to transform it into benzene oxide, which in turn undergoes several enzymatic reactions until it becomes phenol, which then turns into a substance like benzoquinone before being released in the urine from the body. Substances like phenol and benzoquinone, however, which are produced during the metabolism of benzene facilitated by the cytochrome enzyme system, are more toxic than benzene and have the capability to damage the DNA or intracellular proteins. In the end, exposure to benzene produces a number of intermediate toxins during the metabolism process that adversely affect important structures or functions in the cells, such as the genes or enzymes. If the altered gene or enzyme does not play its given role appropriately, various complications leading to disease can occur in the body.

The detoxification mechanism, which uses our body's metabolism, is basically designed to decompose the toxins that infiltrated the body in the liver, and to make them soluble in water, before discharging them through the kidney into urine. The fundamental reason that the human body has come to possess the metabolic mechanism is that the various nutrients from the ingested food need to be broken down and dissolved in water before being carried by the bloodstream to the cells of the tissue that needs nutrients, because it is necessary for the nutrients to be able to be transported in a soluble state in the blood. Then, the nutrient must enter the cell upon its arrival to be used, but as the cell membrane is composed of lipid, nutrients soluble in water cannot enter the cell directly. For this reason, the cell membrane is further equipped with a special messenger protein that facilitates the selective entry of only the necessary nutrients into the cells.

This process may seem a bit complicated, but it can be simply summarized as follows. There is a series of mechanisms by which the ingested foods are metabolized to water-soluble nutrients; at first the bloods are used to carry such nutrients over to the cells, and then the delivered nutrients are transferred to the cells using the transfer protein. The reason for the existence of such various stages in the mechanism may be that it is necessary to dissolve various foreign substances in the blood and to allow only the nutrients needed by the cells to enter them. This is thus a defensive system that prevents unnecessary nutrients and toxins from entering the cells.

If the toxic substance itself is a fat-soluble substance that can be dissolved in the fat component, however, not all of it is metabolized as a water-soluble substance in the liver, and some of it may remain as such in the blood. In this case, the fat-soluble toxins can pass through the cell membranes made of lipid components and can enter the cell. The system for defending against toxins, however, has been elaborately developed by considering the entry of even unnecessary or poisonous substances into the cells, suggesting that some toxins can be treated inside the cell. In other words, there is a special binding protein that can extract poisons out of cells by binding with such toxins, even though they successfully entered the cell, so that they can be discharged via the urine. To recap, the toxin-processing proteins involved in the transformation, conjugation, and excretion of toxins cooperate elaborately with one another to remove the toxins in the body.

The proteins that treat toxins, however, usually do not treat just one but rather several different toxins at the same time. In addition, they do not always exist in a certain amount in our bodies; their amount increases when many toxins come in and grows smaller when there are only a few toxins in the body, thereby ensuring that no unnecessary amount of such proteins remain in the body. It can be seen that this flexibility has been built in our bodies to help them better adapt to the environment and to ensure that the materials that our bodies need are not removed inadvertently from our bodies.

On the other hand, there are some cases in which toxins become more toxic and produce carcinogenic chemicals in the process of converting toxins into water-soluble substances, such as the case of benzene metabolism. Therefore, the role of the proteins binding with toxins is crucial in the process of detoxification. The proteins binding with toxins not only bind with toxins but also remove them and prevent them from binding with the genes or damaging the proteins in our bodies. Therefore, not just the cytochrome enzyme system but also the proteins responsible for the production and regulation of the other proteins that bind with toxins play important roles in directing the detoxification process. For example, Nrf2 is a protein that regulates the antioxidant response when toxins enter a cell.[56] In normal cases, Nrf2 remains inactivated in the cytoplasm, but when the oxidative stress is increased due to the presence of toxins in the cell, it moves into the nucleus and activates a gene that produces detoxifying proteins, such as proteins that bind with toxins. For this reason, the body's overall detoxifying ability can be enhanced by ingesting foods like broccoli, green tea, garlic, and blueberry, all of which work to activate Nrf2.

As toxins can sometimes become more toxic in the process of being transformed to water-soluble substances, the final outcome of toxicity can vary depending on how fast the proteins that bind with the "transformed toxic chemicals" are acting. If the balance between the activity of the transformation of toxins and the activity of the chemical binding proteins is broken, thereby accelerating the water-soluble transformation of toxins to produce more toxic chemicals while diminishing the amount of available proteins binding with the toxic chemicals, it can escalate the toxicity and damage the cells. Therefore, it is essential to know the factors that disrupt the delicate balance between the toxin-transforming proteins and the chemical-binding proteins. The factors that break this balance include alcohol intake, smoking, taking drugs, and aging.[57] For example, the excessive use of acetaminophen can lead to the depletion of the proteins binding with the toxic chemicals and result in increased oxidative stress, inflicting potentially significant damage on the liver cells.[58]

In this case, a drug that produces proteins capable to bind with toxic chemicals should be administered to maintain the balance between the toxin-transforming proteins and the chemical binding proteins, to prevent damage. Therefore, we need to acquire more detailed information about these proteins by monitoring detailed molecular profiles so that we can keep the balance appropriately. To do this, we need to know more about the systems involved in the metabolism of toxins, such as genome, epigenome, proteome, and metabolome, which are constantly changing to work together.

Toxic chemicals are the main cause of oxidative stress

The mitochondria inside the cell produce energy using nutrients and oxygen, which in turn produce reactive oxygen species associated with oxidative stress. These reactive oxygen species are not just discarded but are used as messengers through which the organelles and molecules in the cell exchange information with each other or for neutralizing or reducing the toxins or intermediate toxic chemicals that infiltrated the body from outside or that are produced during the metabolism process, through oxidization. In addition, the activation of cell regulatory genes by the reactive oxygen species has a significant effect on the cell activity, including inducing suicide of cells that have fallen in function. Excessive amounts of reactive oxygen species, however, can cause harmful reactions. For instance, inflammation is triggered by reactive oxygen species activating inflammatory cytokines and chemokines, which are defense mechanisms of our body. When these inflammatory reactions, however, are persistent, they are also very important factors that aggravate chronic diseases such as cardiovascular diseases, diabetes mellitus, and cancer, and the late chronic diseases such as neurodegenerative diseases and autoimmune disorders.

Reactive oxygen species are also produced frequently when toxic chemicals are metabolized by the toxin metabolism system. Oxygen is used to convert lipophilic toxins to water-soluble toxins, producing reactive oxygen species as byproducts in its wake. Therefore, the greater the exposure to toxic chemicals, the greater is the amount of reactive oxygen species produced. Fortunately, there is an antioxidant system in the cell for preventing hazardous oxidative stress responses by an excessive amount of reactive oxygen species, the reason being that the surplus oxygen species can attack the mitochondria in the cells, the DNA or the other molecular structures in the nucleus, or the membrane structures that are made of lipids, such as the cell or nuclear membranes, and then degrade their normal cell functions. On the other hand, these attacks are indiscriminate rather than targeted at a specific molecule or intracellular structure, which is why many different diseases, such as hypertension, diabetes mellitus, heart disease, and cancer, are nonspecifically triggered by deleterious oxidative stress responses.

The antioxidant system was developed back in the days when the amounts of toxins entering the human body from outside were small. Therefore, in today's living environment, in which people are constantly exposed to numerous chemicals, the antioxidant system is considered too old to manage the reactive oxygen species that are being generated in excessive amounts inside the body. In other words, many chemicals produce reactive oxygen species in the human body, causing oxidative stress, which is the central factor that causes or exacerbates various chronic and late chronic diseases. Therefore, efforts are needed to maintain the balance between the production of reactive oxygen species and the activity of antioxidant systems. To reduce the production volume of reactive oxygen species in the body, information on all the chemicals to which we are exposed in our living environment should be available first. Furthermore, we need to know the degree of exposure of ourselves to a specific chemical that contributes to the production of reactive oxygen species. This is because only after detailed information on toxic chemical exposure and its impact is available can we accurately understand the impact of reactive oxygen species and successfully manage them.

At the same time, various efforts to supplement the antioxidant system should be made, such as the consumption of foods containing numerous antioxidants, including vegetables and fruits, or antioxidant vitamins such as vitamin C or E. Only when we know the amount of antioxidants that are present in our bodies or how effectively they function, however, can we accurately reinforce the antioxidant system. Therefore, it is crucially important to evaluate our exposure to toxic chemicals related to the production of reactive oxygen species and to continuously monitor the function of the antioxidant system while reducing the oxidative stress in our bodies, if we aim to prevent and treat chronic and late chronic diseases successfully.

The first step in disease prevention: avoid contact with toxins

If we examine all the toxins to which we are exposed and the resulting responses of the human body, we can become certain of which substances we should avoid, which nutrient we should intake, or which medication we should take. The problem, however, is that it is more likely that various toxic substances produce diverse health effects through the same reaction than it is that specific toxic substances will have a specific health effect. In other words, a considerable number of toxins cause the same reaction, such as oxidative stress or inflammation reaction, and this reaction may lead to various diseases rather than to specific diseases. Therefore, the impact on chronic or late chronic diseases can be eliminated by identifying all the different exposure factors that can affect such response and finding appropriate measures to manage such factors. In the end, it is important to exert efforts to reduce our total exposure to toxic chemicals in our everyday life because various toxins have the same mechanism of action.

The most critical pathway through which toxins enter our body is food. Therefore, the most basic way to prevent the entry of toxins into our body is to eat less. One of the important principles of Ayurvedic medicine, which emerged in India in around 3500 B.C., is "Panchakarma," which is designed to allow the body to rid itself of wastes that have accumulated and lodged in the body.[59] This concept strives to prevent diseases and to obtain mental stability by cleanaing the intestines through the gradual reduction of one's food intake and by practicing meditation and walking. Even today, regaining one's good health and bodily stability by reducing one's food intake is a very important healthcare method. For example, rice contains nutrients like carbohydrates but also heavy metals like cadmium and arsenic, and other environmental hormones (e.g., phthalates) are also introduced through the food containers. Therefore, not only can excessive food intake lead to obesity and various chronic diseases due to overnutrition, but it is one of the important routes through which toxins enter the body in excessive amounts.

In addition, an originally safe raw material may turn into a toxic substance depending on how it is cooked. For example, toxins like heterocyclic amines or polycyclic aromatic hydrocarbons are produced in meat when it is grilled too much.[60] When eating fruits or vegetables, the pesticide residue therein may also enter our body. When these substances enter our body, they undergo a toxin metabolism process for detoxification, and the reactive oxygen species generated in the process can cause inflammation chronically or affect our genes, leading to the development of various chronic diseases, such as diabetes mellitus and cancer. Therefore, to prevent chronic diseases, it is very important not to consume more than the amount of food that our body needs as a source of energy for our daily life activities.

Environmental toxins may enter our bodies not only through the food that we eat but even while we are breathing. Toxins like the heavy metals and polycyclic aromatic hydrocarbons attached to the fine dust in the air and volatile organic chemicals like the toluene and formaldehyde released from wallpapers or floors in the houses or from the new furniture in the office are among the toxins that we frequently encounter in our everyday lives. When these chemicals enter our bodies through our respiratory system, they also can cause oxidative stress via toxic reaction or metabolic processes, which can have a potentially significant impact on various organs such as lung, the cardiovascular system, and the nervous system.

As stated previously, although the toxin metabolism system is basically a defense mechanism that has been equipped to eliminate various toxins, it is not sufficient to safely remove or neutralize all the new chemical substances that have emerged since the Industrial Revolution. Therefore, these chemicals often cause oxidative stress excessively by providing direct toxicity or during the metabolism process. As the human metabolism and antioxidant systems are incapable of fully responding to such toxic chemicals, the best strategy is to avoid chemical substances as much as possible. However, since not all harmful exposures can be avoided in reality, a more realistic and rational strategy is needed.

To provide accurate instructions tailored for each person to avoid hazard from exposure to toxic chemicals, all the external exposure factors, the toxin metabolism system and oxidative stress in the body, the genes and biological response markers, and the current disease status should be evaluated thoroughly. If we reinforce the antioxidant system by using such information and providing appropriate remedies and establish a system that provides feedback by continuously monitoring the biological response, it will be very helpful to prevent the occurrence of chronic and late chronic diseases. To prevent and successfully manage diseases, therefore, a strategy that monitors and reinforces the toxin metabolism and antioxidant systems as well as continuously providing information on various chemical exposures should be taken.

3.3 Improve Your Immunity to Protect Yourself From Intruders

Immunity is the ability to protect oneself from the intrusion of foreign substances by distinguishing oneself from external substances and fighting them. Having evolved over time, the immune system is basically composed of two subsystems: the innate immune system and the adaptive immune system. The innate immune system prevents bacteria or viruses from entering the body using physical barriers like the skin or mucous membrane, and resists infection by secreting substances that cause inflammation or fight microorganisms. The adaptive immune system is composed of humoral and cellular immune response, providing specific reaction to antigen that penetrates the nonspecific immune defense. Among the immune disturbances, allergic diseases can be understood as reactions that try to defend the body excessively, while autoimmune diseases occur due to the failure to distinguish oneself from external pathogens.

Immunity, the ability to protect oneself from external substances

In 1796, having heard that certain cow-milking women did not acquire "smallpox" after suffering from a mild case of "cowpox," Edward Jenner injected the pus of a cowpox patient into a young boy named Jamie Phipps, who worked at a local farm. Jamie suffered from a mild case of cowpox but recovered soon. Two months later, Jenner again injected the pus of a smallpox patient into Jamie and confirmed that he did not acquire smallpox. Although occurring nearly a 1000 years after the development of the Indian smallpox vaccination method, Jenner succeeded in coming up with a much safer inoculation method against smallpox. In fact, Jenner did not conduct his experiments armed with a full understanding of the immune response, and his experiments cannot be considered ethical when judged by today's standards. It cannot be denied, however, that Jenner's experiments led to a remarkable scientific achievement that paved the way for the prevention of infectious disease caused by the pathogens: the development of the first effective vaccination method by capitalizing on the body's immune response to prevent terrible infectious diseases.

Immunity is the ability to protect oneself from the intrusion of foreign substances by distinguishing oneself from external substances and fighting them. Why, however, are so many of the microorganisms that are in a symbiotic relationship with our bodies not regarded as invaders even though they are external substances? This is probably due to our ability to accept symbiotic microorganisms as our own cells or friendly ones through the long process of coevolution. In fact, the body's relationship with symbiotic microorganisms is very important for the formation of immunity functions. For example, the immune system of a fetus is not yet mature; thus, it is necessary to deliver an appropriate stimulus to the fetus's immune system so that the fetus will gain the ability to fight the bacteria or foreign substances that may enter its body later. When an infant is born, it is continuously exposed to microorganisms, starting from those residing in the vagina of its mother to a number of microorganisms normally present in the living environment, all of which play the role of appropriately developing the immune function of the child. The immune cells of a child, while normally staying in contact with the microorganisms in the human body or in the living environment, are being stimulated by the microorganisms to develop the immune function maturely or to develop the ability to distinguish symbiotic microorganisms from infectious agents or toxins. Seen otherwise, it can be said that the stimulations to the child's immune system to develop are obtained through the symbiotic relationship with microorganisms. Thus, microorganisms and their hosts do not only live together but also develop the ability to adapt to each other for a symbiotic relationship with each other.

Immunity is the function of identifying and defeating the microorganisms that are harmful to the human tissues and cells; on the other hand, however, it has been developed in the symbiotic relationship between humans and microorganisms. Some bacteria that once lived in natural environment have adapted to live in the body of an animal as well; therefore, the health of the animal is very important for the bacteria because the animal is a shelter that provides the bacteria with abundant nutrients. From the point of view of animals, they cannot survive without bacteria living in their bodies because they acquire the nutrients from digestion of food by the bacteria. The problem, however, is that even though the majority of bacteria are beneficial to their host animals, some others can cause diseases. Therefore, the need for the ability to distinguish beneficial bacteria from harmful ones has played a role in developing animals' immune system.

In the end, understanding the body's symbiotic relationship with bacteria is very important in coping with immune disturbances. It has been shown through experiments that symbiotic bacteria stimulate lymphocytes to differentiate into the T cells involved in cell immunity, suggesting that bacteria play a very important role in the formation of the immune function.[51] In fact, we are in a symbiotic relationship with most microorganisms in our living environment, and for the pathogens or unfamiliar substances that are unable to form a symbiotic relationship with us but can cause disease, we fight them with our cell immunity or antibody production to neutralize them. In addition, to mobilizing such immune system, the human body also has a defense mechanism that protects itself by producing inflammation-inducing substances that cause inflammation reactions by activating the white blood cells.

To defend the body, the immune system basically starts with the distinction between the self and the other. This is because the human body can determine whether it is safe or dangerous only after distinguishing itself from the other, and if human body determines that it is in danger, defense from the threat posed to our body needs to be activated. As such, immunity is mostly developed to prevent the invasion of microorganisms, and the defense process starts with distinguishing oneself from microorganisms by recognizing specific parts of the microorganisms. For example, we get to notice that bacteria have invaded our body by recognizing antigens like lipopolysaccharides that are present in the cell walls of gram-negative bacteria. It then activates the defense mechanisms, such as cytokines or reactive oxygen species, to initiate the body's defenses against the invading microorganisms.

The immune system or immune response is not fixed but has changed constantly and will keep changing in the future. From the birth of the first organism to the development of immune system of humans, programs that initiate an immune response have continued to evolve. In addition, the relationship with pathogens changes from the birth to the death of individuals during one's life and even the number and proportion of immune cells that play different roles within a single infection period can change as well. In other words, the immune function is not a fixed function but changes constantly depending on the conditions of the pathogen or antigen.[61] In the case of the viruses that people have been infected with for only a short period of time in the history, such as HIV, which causes AIDS, people have yet to establish an adequate immune system against them. Therefore, such viruses compete with the human body intensely, pushing up the infection rate to a very high level, because they have not found a way to coexist with humans. On the other hand, some types of papillomaviruses, which occasionally cause warts, have been in contact with humans for quite a long time and are often found in the skin and mucous membranes of people. In other words, having established a considerable level of coexistence with humans, a certain type of the papilloma virus generally does not cause disease even if it is present in the skin or mucous membrane.[62]

Two-stage immunity shield

Immunity is a mechanism for recognizing and defending the self from other organisms or substances. Having evolved over time, the immune system is basically composed of two subsystems: the innate immune system and the adaptive immune system. The innate immune system prevents bacteria or viruses from entering the body using physical barriers like the skin or mucous membrane and resists infection by secreting substances that cause inflammation or fight microorganisms. It is a nonspecific immune system, with "nonspecific" here meaning staging a general and basic defense rather than coming up with a response specific to the external threat by identifying the nature of the threat.

Examples of the innate immune system include washing out microorganisms or foreign substances with tears or mechanically pushing out microorganisms or foreign substances with bronchial ciliary movement. There is also the method of destroying bacteria with an enzyme like lysozyme contained in the saliva or the method of chemically protecting the body from bacteria by maintaining low pH levels in the stomach or vagina. Meanwhile, the bacteria that are normally found in the intestines prevent other pathogens from entering the body by acting as defensive forces. Leukocytes, which are involved in the inflammatory reaction; histamine, which enlarges the capillaries to promote blood circulation in peripheral tissues; and the chemical compounds that induce inflammatory responses by attracting macrophages, such as prostaglandins, also constitute the non-specific immunity. In particular, leukocytes, which play a central role in non-specific immunity, are mobilized to counter the pathogens coming into the body. The pathogens that are eaten by macrophages among the leukocytes enter the small vesicles, which are generated in the macrophage by phagocytosis, before they are decomposed by the digestive enzymes or killed by the reactive oxygen species. This activity of the macrophages can be seen as a trace of the primitive cell activity of consuming external microorganisms as food, which has come to make up a part of the immune system.

One of the non-specific immune responses is a protein called "interferon" produced by a virus-infected cell at the time of virus invasion. Interferon plays an important role in preventing the synthesis and proliferation of viruses in infected cells. Viruses, unlike bacteria, can hardly be prevented by leukocyte-like defense systems. Therefore, intracellular defense mechanisms called interferon also need to exist to prevent the spread of viruses among cells. On the other hand, when pathogens come in, the body temperature rises, accompanied by fever and chill. As most pathogens are sensitive to temperature, their ability to survive in the body declines when the body temperature rises. It can be concluded, therefore, that fever appears to suppress pathogens to defend against them. Excessive fever, however, is surely dangerous because it can impede the function of the normal cells of the human body, but the infection and the accompanying fever can also be seen as a nonspecific defense mechanism of the body.

The specific immune system defends the body from the pathogens that penetrate the non-specific immune defense, and the external factor that causes a specific immune response is called an "antigen." In fact, adaptive immunity is a specific response to an antigen, and its most important feature is that if you come into contact with an antigen that you have already contacted earlier, a very rapid and powerful immune response occurs. The adaptive immune system can be divided into humoral immune response, which is activated by the B lymphocyte, and cellular immune response, where the T lymphocyte plays an important role. B lymphocytes are made from the stem cells in the bone marrow before they are released into the blood to engage in activity. T lymphocytes are also made in the bone marrow and then enter the thymus before gradually moving out into the blood to engage in activity. The lymphocytes released into the blood identify the external pathogens or foreign substances entering the body from outside by memory and recognition and engage in the activity of neutralizing them.

For example, if pathogens invade the body, the macrophages that captured such pathogens can transmit related pathogen information to the T or B lymphocytes to make them prepare a matching response. The T lymphocytes are then differentiated into cytotoxic T lymphocytes to directly kill the infecting bacteria or viruses, and the B lymphocytes are differentiated into plasma cells, which produce antibodies capable of blocking the invasive microorganisms using the information on such microorganisms. These differentiated T and B lymphocytes not only prevent the activity of invasive bacteria or viruses but also store the pathogen information in the cells; as such, if the same bacteria come back to haunt you in the future, your body will have the ability to block them.

The reason that the adaptive immune system is composed of a cellular immune response as well as a humoral immune response that creates an antibody to an antigen is that there are cases in which an antigen × antibody reaction alone cannot construct a sufficient defense system. For instance, the virus can infect the cells by avoiding the antigen × antibody reaction, and the antibody alone cannot defend the cells that have become independent of the normal cell control, such as the cancer cells. In this case, a cell-mediated immune response is mobilized to remove the virus-infected cells or cancer cells. In other words, T lymphocytes that directly attack and destroy the diseased cells need to be activated in such circumstances. The cell-mediated immune responses also play an important role to prevent infection by fungi and protozoan like amoeba.

As such, the immune system offers two-stage protection. The first shield, a nonspecific defense system, can be regarded as ramparts aimed at blocking out intruders from outside, and the second shield is like a guard that defeats the external intruders that were able to penetrate the ramparts. If an intruder, such as a pathogen, however, is able to penetrate the two-level shields, an infectious disease may occur. On the other hand, if the defense activity against external intruders like microorganisms or foreign substances is too intense or beyond the usual levels, an allergic disease may occur. Meanwhile, an autoimmune disease is a disease in which B or T lymphocytes are activated beyond the usual levels, thus developing immune responses to the person's own cells despite the absence of the usual causes, such as pathogens or cancer cells.

Diseases arising from the inability to distinguish friendly and enemy forces

Although the immune system consists of a defense system that has been structured elaborately over several stages to combat pathogens or foreign substances, the immune system itself may sometimes lose its stability and fail to function normally in the course of the attacks and defenses that happen constantly. Diseases that occur when a symbiotic relationship between the self and the surroundings cannot be established normally or when the symbiotic system is disturbed, thereby causing confusion in the distinction between one's identity and the enemy, are called "immune disturbances," with allergic diseases and autoimmune diseases being examples of such.

Among the immune disturbances, allergic diseases can be understood as reactions that try to defend the body excessively when foreign substances come in or come into contact with the body. Pollen, fungus, and dust are not toxic in their own right and do not cause serious infections, but they can cause severe allergic reactions. Such reactions occur when the system for recognizing the foreign substances in our body and defending the body from them is not yet mature enough or is overly responsive to some stimuli. Diseases like allergic rhinitis, asthma, or atopy are caused by this immature immune system. In short, such reactions occur because of our body's inability to properly control its immune response to external intruders.

In addition, the immune system sometimes attacks the cells by forming antibodies against the person's own cells due to its failure to distinguish itself from external pathogens, which is the reason that autoimmune diseases occur. In other words, the immune cells attack the cells that are functioning normally and impede the function of the attacked cells, resulting in disease development. For example, if an autoantibody formed against a red blood cell destroys the cells by attaching to it, hemolytic anemia occurs; when an autoantibody to an acetylcholine receptor at a junction connecting a nerve and a muscle is generated, the autoantibody will inhibit transmission of nerve signals, resulting in the occurrence of myasthenia gravis. Also, autoantibodies to the thyroid-stimulating hormone receptor cause Hashimoto' thyroiditis, which triggers hypothyroidism because it impedes the production of a sufficient amount of hormones in the thyroid cells.

The cells that make up the human body get together to represent one person, but they are different from one another. As cells, however, share the same sign that they are in the same league, they use this sign to inhibit the activation of the immune system. It is as if they are soldiers in a war relying on markers to distinguish the friendly forces from the enemy forces. If these signs disappear, however, or if the "soldiers" fail to recognize them, the soldiers get confused and could attack the friendly forces. As there are such markers like the human leukocyte antigen on the surfaces of the human cells, it is possible to perceive different cells as belonging to the same person. If such a system that recognizes itself and the other gets confused, however, it may show excessive reactions to external substances or launch an offensive against the person's own normal cells, resulting in the occurrence of an allergic or autoimmune disease.

The immune system has the function of recognizing and preventing dangerous situations that may arise from inside the body as well as the function of recognizing pathogens or other foreign substances and preventing them from entering the body. Therefore, it is a system designed to cope with a variety of internal and external environments to protect our body as well as to prevent entry of foreign substances into the body. As cancer cells do not have markers like the human leukocyte antigen, natural killer cells recognize them as other cells, not as the body's own cells. If an error occurs in the recognition function, the cancer cells cannot be killed, resulting in the occurrence of various types of cancer. In the case of rheumatoid arthritis, on the other hand, it is a disease caused by the body's immune system erroneously recognizing the tissue cells surrounding the joints as "others."

As such, a functional failure of recognition in the immune system can thus cause cancer or autoimmune diseases as well as allergic diseases.[50]

These diseases occur mainly due to the fact that the immune system is not functioning normally; they can also occur, however, because the symbiotic relationship with microorganisms that have made a decisive contribution to the formation of the body's immune system has been broken. The bacteria that are in a symbiotic relationship in the gastrointestinal tract or mucosa have played a critical role in the formation of the human immune system. The bacteria living in the gastrointestinal tract promote the anti-inflammatory response, inhibiting the inflammation triggered by other pathogens, and protect the epithelial cells that make up the epidermis inside the intestines. If the total number or diversity of the bacteria that are living normally in the gastrointestinal tract is reduced, however, this protective role will not work appropriately. The tight junction to provide a protective physical barrier of the intestinal epithelial cells is also very important for immune disturbances. The large intestinal epithelial cells are tightly interconnected, but there is a slight gap between the cells, through which the necessary nutrients are taken and the unnecessary substances are released. However, if the gap is too wide from the beginning or if the intestinal epithelial cells are damaged so that the gap becomes wider, it prompts unnecessary foreign substances to easily enter the body, thereby causing a variety of problems including immune disturbance.

For instance, if the amount of bacteria living normally in the large intestine is reduced due to dietary changes or the use of antibiotics, pathogens or toxic bacteria can multiply to alarming levels. If they do, they will damage the intestinal epithelial cells, widening the gap between the cells. Meanwhile, as the intake of processed foods increases, new chemical substances that had not been recognized in the past may continue to enter the body through the widened gaps of the intestinal epithelial cells. When this happens, the immune system is likely to be disturbed, possibly due to the overload of foreign chemicals. As a result, the ability to distinguish between the self and the other is deteriorated, prompting an immune response to the cells constituting the human body. Inflammatory bowel diseases such as Crohn disease or ulcerative colitis may be caused by the disturbance of the immune system stemming from such a large gap in the large intestinal epithelium, which then leads to the development of an autoimmune reaction.[63]

There are nearly 80 different autoimmune diseases in existence, making such disease type the third most common in advanced countries after cancer and heart disease.[64] It has a wide spectrum, ranging from diseases mainly arising in the joints, such as rheumatoid arthritis, to diseases that appear on the skin, such as vitiligo or scleroderma, and diseases that appear on the thyroid, such as Hashimoto thyroiditis or Grave disease. In particular, type 1 diabetes mellitus was not known as an autoimmune disease in the past, but it is now classified as an autoimmune disease. As such, autoimmune diseases appear as diverse diseases, but sometimes they are difficult to be diagnosed specifically because very similar symptoms frequently overlap to occur, and in some cases, two or more diseases may occur at the same time. This complexity is compounded by the fact that there is no specific relationship between antigen and autoimmune diseases. On the contrary, in the case of autoimmune disorders, various diseases could be manifested by various antigens when the immune system is disturbed.

How to normalize the immune function

It seems simple to distinguish the self from others, but it is in fact not easy because the pathogens that want to survive by invading our body strive to penetrate the surveillance network that distinguishes the self from others and to exert continued efforts to neutralize our defense system. Ultimately, the competition between the defense system and the pathogen, the former to make its surveillance system more elaborate and the latter to try to penetrate such surveillance system all the time, has created the immune system as we know it today, and it is still in the process of changing. Seen this way, the immune system is a defense system for the survival of humans, and the cycle of attack and defense will continue until the body's adaptation to pathogens or foreign matter is fully realized, achieving a peaceful coexistence, which, however, will not happen anytime soon.

The immune system is a very sophisticated system that works based on the principle of balance and harmony. As such, it is difficult to find a simple way to improve the body's overall immune functions. Nevertheless, several factors have been found to have a significant impact on the body's immune function. If you are old, stressed out, or fed poorly, your body's immune system will be weakened. For instance, when influenza is prevalent, it usually does not bring about serious consequences on young adults, but elderly people can be hospitalized or die on account of it. Your immune function is suppressed when you feel stressed, depressed, or anxious, but it improves when you feel better. Moreover, if you exercise regularly or properly, you can recover or maintain your immune function even after you've gotten older. In addition, lack of nutrients and vitamins may cause your body's immune function to deteriorate significantly, but if you ingest nutrients in a balanced manner, your immune function will be restored. It is also helpful to increase your intake of foods loaded with diverse nutrients, such as vegetables, beans, nuts, seeds, and whole grains, which improve the immune function; foods such as yogurt, which enrich the intestinal microbial environment; or foods containing nutrients like omega 3, which suppress inflammation. Other supplements, such as digestive enzymes, which improve the digestive function, or glutamine, which promotes the recovery of the damaged digestive tract, can also help boost the body's immune function. To conclude, various methods, such as physical activity, stress management, proper dietary intake, and some supplements should be utilized to improve the body's immune function.

On the other hand, the vast majority of the current treatments for allergic or autoimmune diseases are steroids or cytotoxic agents. They are medications used to control or modulate the body's immune response, but they should be used with caution because they can suppress the body's normal immune function or be accompanied by serious side effects. Some methods, including immune stimulation therapy or suppression of hypersensitivity reactions, are currently being developed and used as methods for restoring the body's immune ability, but it is difficult to conclude as of this writing that such treatment methods have demonstrated excellent results. Therefore, other methods, such as bone marrow transplantation or cell regeneration, should be developed to selectively control only the pathological immune responses related to disease, without inhibiting the body's normal immune response, or to transform the pathological immune system into a normal immune system. Finally, what is needed to strengthen the body's immune system is not a mere therapeutic approach but a correction of the diverse and multidimensional factors that cause immune disturbances.

It is not simple, however, to pinpoint the factors that affect the immune system, because one hardly sees specific exposure factors causing specific immune disturbances and because various exposure factors in the living environment, including food intake, can exert an impact in a variety of ways. A number of factors, various immune reactions, and various immune disorders are linked in a complex way. That is why it is important to monitor and manage people's living environments during childhood—in particular because the immune system is formed mostly during the period from the fetus to the infant. In addition, external factors and internal responses should be assessed in a timely manner and managed properly because the immune system itself changes over time as one ages. In other words, the incidence of immune disturbances will be reduced only when we get to understand that such diseases are caused by the inability of the body's defense system to function normally against pathogens or foreign substances, and thereafter, when we adopt a comprehensive systems medicine approach to combat such diseases.

3.4 People Should Go Through a Healthy Aging Process

Homo sapiens have ended up with the most complex and superior brain system among all the living organisms on Earth as a result of the growth of their brains' neural networks. However, the incidence of neurodegenerative diseases, such as Alzheimer and Parkinson disease, will increase because neurological function declines with aging. The neurodegenerative diseases are caused by the inability to provide the necessary energy to keep protein arrangement and neuronal activity in the brain, mainly due to the deteriorating function of the mitochondria with aging. In the end, maintaining a healthy symbiotic system with the mitochondria through the overall monitoring of one's lifestyle and environmental exposure and bodily and mental functions can prevent the occurrence of geriatric diseases and frailty and can enable one to enjoy a healthy aging process.

The human brain, the most complex system in the human body

The human brain measures up to 1350 mL by volume, about three times larger than the brain of *Australopithecus*, with a volume of about 400–500 mL. The human brain in fact accounts for only 2% of the body weight, but it accounts for about 20% of the body's total energy consumption, making it an energy-hungry organ. High energy consumption is disadvantageous for survival, but the reason that the human brain has grown so large despite its high energy consumption is that the benefits that humans can gain through their brain activity far outweigh the cost of energy that their brain consumes. Therefore, the force of natural selection worked in the direction of ever-increasing brain size. In particular, the brain volume has begun to increase in earnest since the emergence of our *Homo habilis* ancestors. The brain grew even bigger with the emergence of *H. sapiens*, owing to the use of complex languages, ingestion of diverse foods obtained through the shift from vegetarian to omnivorous diet due to addition of hunting to their food acquisition activity, and competition with one another for more sophisticated technologies related with the fabrication and use of tools. These changes have enabled mankind to equip themselves with the superior recognition and judgment skills required for communication, future planning, and addressing complex problems.[65]

A human organ that enables us to recognize and respond to the surrounding environment, the brain is greatly affected by sociocultural factors. As humans are influenced by complex sociocultural factors, there is a greater likelihood that they will hold the upper ground in the natural selection and respond better to the surroundings when the fine networks of brain cells are more elaborately structured. Moreover, given the complex sociocultural influences, epigenetic programs as well as the natural selection of genes play an important role, transforming the cerebral network of individual brain into a finer neural network. In this sense, the brain of each person is a human organ, but it can also be said to be a network that exchanges influences directly with sociocultural systems. For example, jealousy, desire, or other social behaviors indicate that people are living in relationship with others, or in a sociocultural system, where the social actions performed by people based on their brain's commands affect other people and also the entire community.

The high-level functions of brain, such as cognition, judgment, and conscience, are the results of the activity of the complex system made up of brain cells. However, the human brain activity, which is in a much higher level than the brain activity of other animals, is not really much different in terms of the way it operates from the brain activity of simpler animals. For instance, a mouse also has a brain that operates in the same way; the only difference between the human brain and that of a mouse is the level of complexity of the two. They have the same structure because the basic structure of the neural network system established by natural selection has not changed much. Therefore, we can say there is hardly any difference between the basic structure of the brains of our early ancestors, such as *Australopithecus*, and that of the brain of modern humans. If any, the size of the cortical areas where neurons are densely clustered has increased significantly, and specialized parts, such as those governing the linguistic functions, have emerged. The larger the cortex, however, the greater is the amount of neurons therein and the greater is the amount of neural connections among the brain cells. As quantitative changes generally add up to prompt qualitative changes, *H. sapiens* have ended up with the most complex and superior brain system among all the living organisms on Earth, including primates, as a result of the growth of their brain's neural networks.

The brain has a complex network of 100 billion neurons as well as 100 trillion synapses connecting these. One of the most interesting aspects of how the human brain system works, however, is that there are no core parts in the brain through which information is processed to perform recognition, judgment, and planning. Rather, the information processing domains, each performing a different function, are distributed all over the different cortices, and information is exchanged among such domains. Therefore, the information is not processed according to hierarchical order but in a parallel and interrelated manner.[66] As information is processed across the distributed network in a parallel way rather than in a hierarchical manner, the brain has potential capacity to compensate for some damages inflicted on the system. For example, people often recover from severe cases of brain damage, such as a stroke, and regain their previous cognitive abilities, the reason being that the brain system has evolved in such a way as to increase its complexity in a parallel and interrelated manner according to the need to respond to the surrounding environment.

Why do our excellent brains fall to neurodegenerative diseases as we age?

Globally, the elderly population aged over 60 is expected to jump from 900 million in 2015 to over 2 billion by 2050. In particular, the number of super-aged people (over 80) is expected to increase more than threefold, from 120 million in 2015 to 430 million by 2050. In addition, the world will enter the full-fledged era of the elderly in 2030 as the number of elderly people then is expected to exceed the total number of children.[67]

As aging is inevitably accompanied by the deterioration of the neurological function, the incidence of neurodegenerative diseases will increase if the elderly population grows larger. The typical neurodegenerative diseases are Alzheimer and Parkinson diseases, which are characterized by the degenerative development of the central nervous system. Alzheimer disease is the leading cause of dementia. Currently, 30 million people worldwide are suffering from the disease. The number of Alzheimer disease patients will increase with the corresponding increase in the elderly population, and by 2050, it is feared that 1 of 85 people will have the disease. After Alzheimer disease, Parkinson disease is the second most common neurodegenerative disease. About 1% of the population aged over 60 have Parkinson disease, but among elderly people aged 80, the percentage is higher: 4%. The reason that the incidences of Alzheimer and Parkinson diseases are feared to increase in the future is that age is the most important factor for the development of neurodegenerative diseases. In other words, neurodegenerative diseases occur more frequently as people age.

Alzheimer disease, which is characterized by memory loss and cognitive impairment, or Parkinson disease, which impedes movement or the motor function, comes about as the proteins normally present in the brain start to aggregate. In Alzheimer disease, amyloid β and τ proteins, and in Parkinson disease, proteins called "α-synuclein," become entangled, causing disease.[68] However, these proteins are not known to have direct role for the function of neurons. In other words, the proteins that normally exist in the brain, although the function is not clear, aggregate to cause disease, suggesting that our brain is using considerable energy to produce and maintain the proteins, and finally can have a neurodegenerative disease called Alzheimer disease or Parkinson disease because of aggregation of the proteins. Why then did humans end up having such proteins that aggregate with one another and cause neurodegenerative diseases? Perhaps these proteins have played an important role in making the brain work more efficiently even though they do not play a direct role in the function of the neurons. Researches have shown that the tau protein and α-synuclein play some important roles in the normal development of the brain or in regulating the release of neurotransmitter such as dopamine.[69,70] Why then do changes occur in the tau protein or α-synuclein, causing Alzheimer or Parkinson disease?

A protein is a three-dimensional structure with some 300 amino acids connected with one another. Proteins can perform their intended functions only when they are folded at precise angles to form a three-dimensional structure when they are produced. Perhaps the τ protein or α-synuclein in the brain may play the role of transmitting neuronal signals using a three-dimensional structure. It can be likened to a semiconductor circuit board, which must have a precise three-dimensional structure so that signals can be transmitted along the circuit and so that it can perform its intended functions. Proteins with this three-dimensional structure may be linked with one another to create a space where the brain can perform complex functions. Probably considerable energy is required, however, to transmit information smoothly while keeping the three-dimensional structure of proteins aligned at regular intervals.

The energy required for such action comes from the mitochondria in the brain cells. The problem is that the mitochondrial function becomes weaker as getting older, and the energy required for maintaining the three-dimensional structure of proteins becomes diminished, creating conformational change of the proteins. In addition, the proteins that were not folded correctly in the cell should be removed, but the ability to remove such proteins gets impaired when the mitochondrial function starts to fall. As a result, the proteins forming a three-dimensional structure may end up being unable to support the array and begin to aggregate with one another. In a word, they are collapsing because of energy loss!

Neurons, which perform the most fundamental neuronal function, gradually disappear as the function of the mitochondria in the brain cells deteriorates and the proteins get aggregated. It is unclear if the aggregated protein is the direct cause of the death of the neurons, but many studies report that the protein itself turns toxic when aggregated. In Alzheimer disease, the number of neurons in the brain cortex decreases, causing memory loss, while in Parkinson disease, the number of neurons that produce dopamine, a neurotransmitter, is reduced, causing a problem in the movement and posture. Thus, we can say that neurodegenerative diseases, known to be manifested as cognitive function disorder such as memory loss or motor disorders including movement dysfunction, are caused by the inability to provide the necessary energy to keep protein arrangement and neuronal activity in the brain owing to the deteriorating function of the mitochondria with aging.

Aging, the price to pay for youth

The increase in life expectancy does not necessarily mean that all people will enjoy good health as they age. There are many people who lose their vitality and who experience difficulty living a normal life as they get older. The condition in which the body weight is reduced, one gets easily tired, and one's physical strength is lessened with age is called "frailty."[71] Frailty occurs in 14% of the population aged over 65 and in over 30% of those aged over 85.[72] Accordingly, a significant increase in the elderly population means that the number of people in a frailty state will increase significantly.

The cause of frailty is that major organs like the brain, heart, and muscles begin to fail significantly with aging. Aging means that people begin to struggle with the degeneration of their cartilage or ligaments, with a decrease in the elasticity of their skin and blood vessels, and with a hardening of the lens cortex in their eyes. It is basically a phenomenon, however, in which the living cells that have survived for a long time gradually lose their respective functions. Although these changes do not directly lead to death, some elderly people become too fragile to live a normal life.

The cells in the human body that are no longer differentiated or regenerated include those constituting the nervous organs (e.g., the brain) and those constituting the muscles of the heart, arms, and legs. Called "postmitotic cells," they are rarely replaced by new cells when they get old or damaged. If the cells damaged in the aging process are not replaced with new ones, they will die out, decreasing the overall number of viable cells or they are still alive with damage, increasing the number of nonfunctioning cells, thereby causing the functions of the brain, heart, and muscles to deteriorate. In fact, there are stem cells in the brain, heart, and muscles that are capable of producing new cells, but only a limited level of cell regeneration is being realized in such organs compared to the other human organs. Therefore, a system that processes the residual byproducts formed in the cell for renewal of the cell is necessary because the vast majority of postmitotic cells are not replaced by new cells. For example, a self-digesting system called "autophagy" can help sustain the cell vitality to a certain extent by recycling the old parts of the cell. As such autophagy system cannot be as complete as the replacement of the cell itself (e.g., cell regeneration), however, it cannot completely prevent the cellular function from deteriorating with age.

Cells have a very efficient system that uses oxygen to oxidize nutrients such as sugars so as to get energy from them, and the factories that drive this system are the mitochondria within the cells. The Krebs cycle of the mitochondria is the key pathway to producing ATP molecules, which are used as energy sources for cellular activities like biosynthesis or cell division. Although the mitochondria is a highly sophisticated and highly efficient system, if it strays outside the balance in receiving oxygen and nutrients and in producing energy, it will have a hard time operating normally. Particularly, when supply of oxygen and nutrients to mitochondria is more than required for energy production, the amounts of reactive oxygen species and remaining sugars from the energy production process build up as byproducts within the cells. These byproducts again interfere with the normal functions of the cells through their attachment to important cell constituents like the DNA, and the cells will lose their functions and eventually die. The cells of the brain, heart, and muscles, however, cannot produce new cells to replace the dead cells and therefore gradually become ineffective. This is the fundamental process of the aging, and the severe impairment of the cellular function is to bring the state of frailty.

It is known that the "hydra," a primitive animal, has an unlimited lifespan or does not grow old. The hydra has no postmitotic cells, and all of its cells are to be differentiated and are then replaced with new ones. Why then do the cells of organisms higher than the hydra, especially the human brain or cardiac muscle cells, become postmitotic cells that are not replaced with new ones? Perhaps they have evolved in such direction through natural selection because that is more advantageous than being replaced. In fact, it is easy to adapt to an environment when the neurons in the brain retain the information on it in their old memory. When the cells are replaced with new ones, the memory is also lost, along with the ability to adapt to an environment. Moreover, as the cardiac muscle cells are tightly connected with one another to enable uninterrupted contraction and relaxation, they do not have enough time to die and to be replaced with new cells because if the cardiac function is interrupted in the process of cell replacement, it can easily escalate into a fatal situation. For this reason, the cells of the important bodily organs, such as the brain and the heart, turn into postmitotic cells. Therefore, the lifespan of humans depends on how long the postmitotic cells of these organs can survive. To conclude, the human lifespan is not infinite but finite because the human body aims to increase its adaptability to an environment and then push up its survivability when it is still young. Therefore, aging can be considered a price to pay for securing survivability during one's youth.

Frailty, which can be considered a state in which one's muscular strength is severely weakened and enjoyment of a normal life is impeded, is not clearly distinguishable from a disease because frail people are prone to fracturing their bones when they fall, cannot move well, are dependent, get sick easily, and may therefore succumb to death sooner. For this reason, considerable medical care should be provided to frail people, which will of course increase the burden on the society. The future medical costs would be much greater than the current medical costs if we have more frail people in the future. If we, however, successfully cope with aging with appropriate measures before a serious mental or physical impairment arises, we can prevent aging from escalating into a frail state to a considerable extent. What then would such appropriate measures be?

How to ensure a healthy aging process

One of the most important purposes of human survival is to transmit genes to one's offspring, thereby ensuring the continuation of humankind. There is another item, however, that is transmitted to one's descendants on top of the genes composed of DNA and the gene operation programs. The mitochondria in the cytoplasm, which is located outside the nucleus of the oocyte in the female body, are also transmitted to the offspring. As described earlier, 2 billion years ago, independent living organisms such as bacteria and archaea undergo fusion together and one of them became mitochondria, thereby establishing a symbiotic system. It is this symbiotic system of the genes and mitochondria that is being transmitted to the offspring.[73] The genes play a role in creating the structure and function of the human body, while the mitochondria produce 95% of the energy required for this structure to work. In other words, we depend almost entirely on the mitochondria to gain the energy we need to engage in activities.

The fundamental reason that functional deterioration occurs as people age is that the mitochondria produce less energy. The reason for this is that having delivered genes to the offspring along with the mitochondria, there is less reason for the body to keep up activity. The organ that is most affected by the reduced energy production by the mitochondria is the brain. The top three energy consumers among our organs are the brain, heart, and muscle. As the number of mitochondria in the cardiac and muscle cells is very high, these cells are not affected seriously by the deteriorating function of the mitochondria, but the brain is extremely sensitive to the deteriorating function of the mitochondria as the neurons in the brain consume the most energy even as the number of mitochondria in them is relatively small. The reason of the relatively small mitochondrial number in the brain might be that as the brain cells are sensitive to heat, and so their function is degraded when the temperature rises, there may not be as much energy-producing mitochondria in the brain cells as in the heart or muscle cells.[74] Therefore, when the mitochondrial function is degraded as people age, neurodegenerative diseases of the brain emerge as the most important diseases.

Meanwhile, mitochondria have mitochondrial DNA, "mtDNA," apart from the DNA in the nucleus. When the oxidative stress increases, however, due to the presence of external chemical substances or to aging, the incidence of mtDNA mutation begins to increase. The mutated mitochondria fail to function properly and produce reactive oxygen species for themselves, which in turn degrade the function of the neighboring mitochondria. In this case, the reactive oxygen species also act as signaling agents that degrade the function of all mitochondria in the cell.

Mitochondria, in fact, seem to be independent organisms inside the cell, which change their own shapes and sizes, and therefore their number changes as they are connected with or separated from one another. The reason is that they have independent genetic information in the cell, and carry out independent activities, such as fusion or fission, by putting together or separating the membranes surrounding one another. While performing such fusion and fission activities, they remove their aged or ineffective parts and use these as nutrients to sustain the health of the cells. Such fusion and fission activity among the mitochondria, however, decrease with age. When the fusion and fission activities are reduced, the activity of removing the mutated mitochondria is also reduced, resulting in the increase of the mutated mitochondria within the cell and the further degradation of the mitochondrial function.

Environmental pollutants and chemicals also play a role in mitigating the function of the mitochondria. As these substances are not familiar, but new to humans, our bodies, when exposed to these substances, consider them foreign matter attacking our bodies, producing reactive oxygen species or causing inflammatory reactions by activating the inflammatory cells. Perceiving this as a dangerous situation, the mitochondria try to reduce their number or reserve energy sources by lowering their function and then reducing their normal energy use, suggesting that they try to increase the person's chances of survival by minimizing the risk of active life and increasing the body's energy reserve. The reserved energy source is either present as sugar in the blood or accumulated in the fat cells. As such, exposure to environmental pollutants or chemicals increases the body's insulin resistance and the risk of diabetes mellitus or obesity. Moreover, when diabetes mellitus or obesity occurs, it can lead to many other chronic diseases, including cardiovascular disease. As the functional deterioration due to aging or neurodegenerative diseases is further exacerbated in the presence of such chronic diseases, we can say that environmental pollutants accelerate the impairment of health caused by aging by degrading the mitochondrial functions.

As a result, the increasing occurrence of geriatric illness and frailty in the course of aging can be deemed to be stemming from the degraded function of the mitochondria. Therefore, to slow down aging and maintain a healthy elderly life, it is necessary to maintain the activities and functions of the mitochondria. On the other hand, exercise or reduction of calorie intake is known to promote mitochondrial fusion and fission activity.[75] When mitochondrial fusion and fission become active, the activity of exchange of metabolites between the mitochondria, and the removal of the mtDNA mutations in the cells is also increased. Therefore, we can get away from the dark shadows that aging can bring by exercising, restricting calorie intake, and avoiding exposure to harmful chemicals, as well as by ingesting nutrients like vitamin D, folic acid, vitamin B6, vitamin B12, and coenzyme Q10, which are known to activate the mitochondrial function, and by eating fresh vegetables and nuts as well as foods containing large amounts of antioxidants. Aging is a natural phenomenon involving various stages, from birth to growth and development, and eventually death, but it can also be regarded as a factor to exert an influence on the activation of the mitochondria. In the end, maintaining a healthy symbiotic system with the mitochondria through the overall monitoring of one's lifestyle and environmental exposure and bodily and mental functions can prevent the occurrence of geriatric diseases and frailty and can enable one to enjoy a healthy aging process.

3.5 Functions of the Human Body Will Be Strengthened

Regeneration is a way of restoring one's function by creating new intracellular organelles, whole cells, or tissues to maintain the function and structure of the damaged or lost organs. However, the regeneration capacity of humans is much lower compared with the capacity of primitive organisms less evolved than amphibians. On the other hand, even in humans, stem cells are already in the tissues of a person before being activated when the tissue is damaged and needs to be restored. With technology development, the diseases that have not yet been conquered can be cured using stem cells in the future. In addition, genetic manipulation, one of the enhancement tools, can change the human genes to improve the normal function of the human body and to prevent the aging process.

Regeneration, another recovery mechanism of our body

Immune reaction or detoxification response is a defense mechanism that prevents our bodies from being damaged by germs or toxins invading from outside. On the other hand, "regeneration" signifies a mechanism for regaining the original function through the creation of a new cell or tissue when damage or loss of function results in the failure to play a given role. In other words, regeneration is a way of restoring one's function by creating new organelles such as intracellular DNA or mitochondria, or whole cells or tissues, to maintain the function and structure of the damaged or lost organs.

Regeneration is the ability to recover that is shared by all organisms, from microorganisms to humans. The phenomenon of regeneration, however, manifests very differently depending on the organism. Even when most of their body was lost, flatworms, such as planarians, can regenerate wholly from the rest of the body by growing from a part of the body, such as the head, tail, or another part of the body. The tadpoles and salamanders also have the ability to regenerate their brains and eyes as well as their legs and tails, albeit not up to the level of the planarian regeneration. This ability, however, is not observed often in birds or mammals. In the case of humans, the liver or skin tissue can be regenerated to a certain extent, but most tissues are fairly limited in their ability to regenerate.

The regeneration of an organism can be regarded as a repetition of the process in which life is born and it grows. For example, when a part of the body is damaged and regenerated, the skin tissue encases the damaged part at first, and the various cells of different functional capacity in the corresponding tissue proliferate in the damaged part. At the same time, an organic connection is made among the proliferated cells, completing the regeneration process. Although the regenerative capacity varies significantly by organism, the phenomenon of regeneration can be said to be similar across all living organisms in the sense that the method learned long time ago while constructing multicellular organisms through the linking of multiple cells is applied in the process.

As the degree of tissue regeneration varies significantly by organism, however, it can be said that the degree of the regenerative ability is established for each organism through the process of natural selection. In other words, if the benefit gained from the recovery of a function through the regeneration of the tissues is larger than the cost of the regeneration, the regenerative ability of the tissue will be preserved. Lizards often lose their tail or get it damaged, and the reason that their tail is regenerated after the loss or damage is that the benefit of regenerating it is much greater than the energy consumed when regenerating it. On the other hand, as the gain from maintaining the function of the tail is not great compared with the cost of its regeneration in the case of rodents like rats, the rodents' tail cannot be regenerated when cut off.

In the case of the hydra or planarian, however, which has the ability to regenerate very quickly when its tissue is lost, it is difficult to explain the regeneration through the natural selection process alone. The reason for this is that as genetic mutation rarely occurs in the case of asexual reproduction of organisms like hydra or planarian, it cannot be said that the pressure of natural selection, in which genetic mutation plays a major role in evolution, was great. As the regeneration process is similar to the growth process in the case of the hydra or planarian, it can be said that the ability was not obtained through the natural selection process but rather that the process of growth and development is being utilized repeatedly. In other words, the hydra and the planarian have the elements necessary for growth and development, and if necessary, they just reuse these to prompt regeneration.

On the other hand, as the tissue regenerative capacity varies depending on the life cycle stage, how similar to the original the regenerated tissue is depends on the stage of development of the organism at the time when the injury afflicted.[76] In general, in very young age, the tissue can be regenerated quite smoothly, making it possible to regenerate it almost to the level of the original tissue in many cases.[77] As getting older, however, the tissue regeneration capability deteriorates and it is very difficult to regenerate the tissue back to the level of the original tissues.

Limited human regeneration capability

After the prokaryotic cells evolved to eukaryotic cells and then to multicellular organisms, the cells of the multicellular organisms were able to further develop into cells with different characteristics. They also acquired an ability to produce stem cells with the same characteristics as the mother cells through somatic cell division. Derived from the stem cells, some cells with specific functions are formed, and these cells are connected with one another to form an entire organism. In other words, stem cells divide themselves and differentiate into the end organ cells with different characteristics. However, even though the differentiated cells continue to divide and produce the end organ tissues with specific functions, some of the original stem cells do not disappear and remain in the end organ tissues, retaining a potential to regenerate when necessary.

In fact, as both plants and animals have been through the evolutionary process from prokaryotic to eukaryotic cells before branching out into the two different kinds of multicellular organisms, the stem cells, which play a major role in the development of multicellular organisms, have such a regenerative capacity regardless of whether they belong to a plant or to an animal. Some plants have the capability to grow to an entire individual organism even if there is only a single cell, while others can grow to an entire individual only when a certain amount of tissues are present. However, the degree of regeneration by the stem cells varies among organisms. For instance, the number of stem cells of humans is relatively smaller than that of salamanders, and the human stem cells are not so active that regeneration capacity of humans is much lower compared with the capacity of salamanders.[78] In general, whole tissue regeneration, which is commonly observed in primitive organisms less evolved than amphibians, is rarely observed in more advanced, higher animals. Perhaps this may be because when such animals are faced with the loss of some tissues, it is more effective for the survival of their species to use energy to sustain their life or to produce offspring than to use energy to regenerate the lost tissues. In the case of mammals, it can be said that there is no room for supplying additional energy to regenerate tissues to the necessary energy required to sustain life in the damaged state.

The regeneration capacity also depends on the organ where such stem cells are located so that there is a difference between the human organs in terms of stem cell regenerative capability; the level of stem cell activity in the brain or in the heart is lower compared with that in the skin or in the gastrointestinal tract. The reason is that the gain obtained by regenerating the cells through the use of the stem cells in the organs such as the brain or heart does not outweigh the gain obtained by maintaining the status quo.

On the other hand, mammals lose their ability to regenerate as they get older as the functions of their nerves, muscles, skeletons, and blood vessels deteriorate with aging. In fact, aging is strongly associated with the decline of a person's regenerative capacity. In other words, the phenomenon of aging can be attributed to the inability to maintain the structure or function of the tissue due to deteriorating regenerative ability. Therefore, if the regenerative capability can be maintained or restored, it can prevent aging and lead to an increased lifespan. Aging, however, cannot be solved simply by increasing the regenerative ability, the reason being that it is necessary to pay a considerable price to improve one's regenerative ability. For example, p53 is a protein that regulates the cellular cycle and inhibits cancer development. Mice with hyperactivated p53 experience an inhibition in the growth and differentiation of their cells, resulting in poor regeneration of tissues and accelerated aging. On the other hand, mice with decreased p53 activity have a higher regenerative capacity and a slower aging rate than those with normal p53 activity. When the function of p53 declines, however, the regenerative ability enhanced as such is, unfortunately, accompanied by an increasing occurrence of cancer.[79]

Mammalian animals reproduce actively and proliferate with ease in general, but their lifespan is limited and the regenerative capacity of their tissue is relatively low. On the other hand, creatures with a high regenerative capacity, such as planarians and hydras, seem to have no limitations in lifespan, suggesting that the capacity for regeneration is different between animals with limited and unlimited lifespan. Why is there such an interesting relationship between regenerative capacity and lifespan? Since the genetic information in planarians and hydras is not renewed through the reproductive process as they undergo asexual reproduction, individual organisms can live long, but their adaptability or evolutionary ability is inferior to those of other advanced animals. On the other hand, species that have achieved remarkable evolution through adaptation to the environment, such as mammals, have adopted a method of renewing their genes with sexual reproduction, thereby ending up with the strategy of maintaining their species through offspring rather than enjoying eternal lives as individuals. Consequently, as the ability to regenerate an individual was not necessarily a priority for them, individual organisms did not have to be sustained, thereby limiting their lifespan. Ultimately, the reason why humans are born with a limited regenerative capability is because the sustenance of the species is more important than the life of an individual.

It is not, however, that even the human race, which occupies the top rung of the evolutionary ladder, does not have the ability to regenerate at all. At the cellular level, the gastrointestinal and skin epithelium cells are constantly being destroyed and eliminated before being replaced with new cells to maintain the functions of the gastrointestinal tract and the skin. Moreover, when one is wounded, we observe that new flesh grows and the wound is healed. The skin cells of our body change completely into new cells after 7 days, and the bone cells change into new bone cells after about 7 years. In the case of very young children, their fingertips often regrow and can be restored to their original shape when cut off. Even in the case of adults, their liver tissues will regrow after being surgically removed partially. In other words, even in humans, stem cells are already in the tissues of a person before being activated in earnest when the tissue is damaged and needs to be restored.

Stem cell therapy capitalizing on its regenerative capability

Prometheus is a Greek mythical figure symbolizing regeneration. Suffering Zeus's wrath as he helped mankind step into the world of civilization by bringing fire to humans, he was tied to a rock and punished, with his liver being pecked daily by an eagle. He did not die as his liver regenerated every night, but this meant that he was punished forever as the eagle kept pecking at his liver. In fact, it could not have been known in the era when the Greek mythology was created that the human liver tissues possess a powerful regenerative capability, but it can be assumed that the people in the past believed that they could have an eternal life and not die if the damaged parts of their bodies were regenerated.

In Goethe's novel *Faust,* there is a scene in which Faust, an extraordinary human, uses alchemy to mix various substances and make a living organism named "homunculus," resembling a human, suggesting that today's cloning, genetic manipulation, and stem cell renewal techniques are modern incarnations of Faust's technology. In fact, Goethe, a European who lived in the Enlightenment Era in the 18th/19th centuries, expressed his desire to create living things by using various materials, just as God created Adam with earth. This aspiration is now being applied to the treatment of patients, leading to efforts to regenerate the human cells and tissues.

In the case of humans, however, the cells responsible mainly for the regeneration function are not pluripotent embryonic stem cells with the ability to produce all organs but adult stem cells with a limited regenerative capacity applicable only to the end organs. The skin adult stem cells cannot produce muscles and the muscle stem cells cannot produce skin. Thus, a single adult stem cell cannot produce any complex part of the body, such as arms or legs, which contain multiple organ systems interconnected with others in a complex manner. Although embryonic stem cells are pluripotent, however, they are not superior in all respects to the adult stem cells. As the embryonic stem cells have the ability to produce all human organs, they can grow out of the control of the neighboring cells and therefore manifest themselves as diseases like cancer.

One of the currently successful therapies among the various stem cell treatments is the removal of hematologic cancer cells via chemotherapy and the transplantation of "autologous hematopoietic stem cells" into the bone marrow. It has the advantage of preventing cancer cells from being implanted together with bone marrow transplantation in the case of "autologous bone marrow" transplantation, thereby preventing cancer from recurring. Similarly, if adult stem cells can be obtained and transplanted again in this manner, it will be possible to regenerate lost tissues using stem cells after removing blood cancer or other cancers in the tissues with surgery or radiotherapy, brightening the future of cancer treatment.

Although the embryonic stem cells do not play a major role in the normal regeneration process of the human body, they have a great potential, the pluripotency, suggesting that they can develop into any cell. Therefore, if the embryonic stem cells or the cells immediately derived from them are selectively obtained and used, it will be useful for replacing or restoring tissues, particularly composed of various types of cells. It can be used to treat patients with diabetes mellitus, heart disease, Alzheimer disease, Parkinson disease, or spinal cord injury. In the case of neurodegenerative diseases, in particular, if the damaged neurons can be replaced using the cells that are derived from the embryonic stem cells, which can differentiate into neurons, humankind will be able to find a way to treat neurodegenerative diseases, which are emerging as difficult-to-treat diseases following the age of chronic diseases.

Let's take Parkinson disease as an example. Parkinson disease is characterized as a constantly falling ability to regulate motion, which makes it difficult for the person to walk, makes the muscles stiff, and impedes the initiation of any action. These symptoms occur because the neurons in the hypothalamus of the brain are dying, thereby hindering the production of neurotransmitters called "dopamine" and rendering the person unable to control his or her body movements. Currently, levodopa is being used as a remedy for Parkinson disease because the brain can convert it to dopamine. Levodopa, however, cannot cure Parkinson disease because its long-term use decreases its efficacy and causes various side effects. In the end, Parkinson disease will not be curable until it becomes possible to ensure that the neurons that produce dopamine will not die, or that the dead cells can be renewed using the cells derived from the adult or embryonic stem cells. Again, this means that stem cell therapy will be one of the available options for successful future treatment of neurodegenerative diseases like Parkinson disease.

Stroke is caused by the clogging or bursting of the blood vessels, which results in inability to supply oxygen and nutrients to the brain cells. As stroke is associated with atherosclerosis, hypertension, or heart disease, if the risk factors associated with these chronic diseases are managed well, stroke can be effectively prevented. On the other hand, as people's life expectancy keeps increasing, stroke is expected not to decrease because the blood vessels themselves undergo the process of aging. When a stroke occurs, it was not just the brain cells, such as the neurons or dendritic cells in the brain that are damaged and then killed; all the other cells in the area are also affected. Therefore, it may be necessary to regenerate all the cells in the affected area using embryonic stem cells, which are pluripotent to become various cells instead of adult stem cells. If embryonic stem cell therapy is successfully done and the brain cells are reconnected to perform their normal function again, we can then say that stroke can also be cured. As such, the diseases that have not yet been conquered can be cured using stem cells in the future. In stroke treatment, however, the cells that are not necessary for the brain, such as the bones and muscles, should not be regenerated in the brain, and techniques capable of preventing the cells from spreading out of control, such as in cancer, are also required.

Let's look at another aspect related to stem cell therapy. Stem cells can be transplanted to prevent tissue aging or death to some extent, but if the original factors that prompted the tissue to undergo aging or pathological development are left unchanged, the newly implanted stem cells may not be able to function to their fullest potential. Therefore, it is also necessary to correct the factors affecting the cells or tissues together with the application of stem cell therapy, and to employ measures that can activate the cells, such as exercise and appropriate caloric intake. In addition, the functional degradation of the mitochondria leads to a decrease in the body's regenerative capacity. In particular, when mutations accumulate in the mitochondrial genes of the stem cells, the stem cells themselves fail to function properly, resulting in decreased regenerative capacity and eventual aging.[80] Thus, methods that can prevent or mitigate the mutation of the mitochondrial genes can also be adopted as a way to enhance the body's regenerative capacity. To conclude, even if stem cells are transplanted and tissues are regenerated, it is still necessary to manage the various factors to prevent aging or functional deterioration.

Enhancement of the human function and gene therapy

Mankind's desire to increase his longevity, enhance his physical functions, and improve his cognitive ability is consistent with his goal of increased survival by better adapting to a given environment. What makes humans so different from other animals is that they have increased their adaptability to the environment by using various tools. Our ancestors carved simple tools like hand axes out of stones, and used fire to illuminate the night and cook meals, which increased their survivability. They also made their homes and got dressed, and therefore escaped from threats like wild beasts and the cold weather. Moreover, beginning about 10,000 years ago, they established a civilized society through farming and herding. In fact, human beings cannot be said to have a superior ability to survive in natural environments when judged by their physical strength alone. They have increased their survivability by using various tools to overcome their limited physical abilities, which in a sense signifies a process of strengthening their bodily functions. When the respective productivity levels of the contemporary man and the hunter-gatherers are compared, the productivity of the modern man can be said to be incomparably greater, but this is mainly due to the difference in the tools that they used rather than to the difference in their physical ability. In the end, the tools that we use are enhancement tools designed to boost our abilities.

Among these, clothing, shoes, and eyeglasses have brought direct enhancement to the human body. Clothing played a decisive role in the *H. sapiens'* triumph over the Neanderthals because of their upper hand ability enduring the cold continental climate, resulting in their survival as the only species of hominid.[81] Footwear contributed to the expansion of civilization by accelerating the speed of movement and warfare, whereas eyeglasses have accelerated the development of civilization by allowing a greater number of people to participate in learning and capacity building. Today, we are not just wearing and using tools like clothing, shoes, and eyeglasses but are also using a variety of other tools to improve our bodily functions and overcome disease and weakness, thereby ultimately trying to live beyond the limits of our lifespan.

Let's think more about eyeglasses. Eyeglasses are obvious human ability enhancement tools because they make it easier for us to see and read when our eyesight fails to perform up to the desirable levels. Of course, there are many people who do not need eyeglasses, but eyeglasses are essential tools for people who cannot see well because their vision is failing. Sooner or later, visual enhancement tools will develop even further, and we will not only be able to receive and view information about whatever and from anywhere through computer-caliber contact lenses but will also be equipped with an ability to see microscopic environments and also far distances, as if we wore microscopes and telescopes at the same time, as well as with an ability to see at night. As these devices evolve further, they can be mounted on the lens of our eyeglasses or can be made to replace the lens of eyeball altogether, to be integrated into the body. The enhanced functions can then be recognized as a natural part of our bodily function. In addition, to such visual or information gathering enhancement devices, three-dimensional printers can be used to make skeletons made of alloy, which is stronger than the human bone, whereas artificial hearts and kidneys can take over the role of blood circulation and waste excretion. Furthermore, the information processing device implanted inside the brain may also bring about a remarkable improvement in the memory and cognitive function.

Gene therapy refers to a method of treating diseases by manipulating the genes of a patient suffering from a disease instead of treating it with drugs or surgery. It is a treatment that turns a gene that is not playing its normal role because of mutation into a healthy gene, or that puts in a new gene so that it can play a role in treating the disease. One of these genetic manipulation methods is the CRISPR system.[82] A defense mechanism originally conceived by bacteria to prevent an attack from a virus, the CRISPR system cuts out the unnecessary genes when they come into the genes. Using such a system that can cut and paste specific genes, humans can manipulate genes and construct the gene structures as they want.

Genetic manipulation can develop into genetic enhancement, which changes the human genes to prevent the aging process, well beyond the level of simply treating a disease, or to improve the normal function of the human body. As expected, gene enhancement can be used to improve physical fitness or ability, or to enhance mental abilities such as memory or intelligence. This enhancement can be achieved by changing the somatic genes, but it can also be done by changing the gene of the germ cell. In the former case, a change occurs only in the person who received the gene-enhancement therapy, but in the latter case, the changes in the gene may be passed down to the offspring. Genetic enhancement passed down to the offspring, however, can be a critical challenge for mankind. This is because it signifies that medical science and its practice can not only prevent and cure disease, but also to create an entirely new human being.

Mankind will be one step closer to the end of disease through medical and technological advances like tissue regeneration or genetic manipulation. The fact that we have developed disease prevention and treatment techniques, however, does not necessarily mark the end of disease. Disease can be ended only when we have managed to solve the so many challenging problems that the development of medical technology can bring, and when we have come up with a medical system capable of realizing such medical technology and a good strategy to end disease. Therefore, in Chapter 4, the medical system and strategic methods that can bring about the end of disease will be further discussed.

Chapter 4

Method of Ending Disease and the Future Medical System

4.1 Systems and Precision Medicine, the Shortcut for Ending Disease

The characteristics of disease changed with the arrival of the age of chronic diseases. Now, we observe a spectrum of diverse diseases that cannot be properly managed with a simple strategy developed in the age of infectious diseases. Because our living environment changes so quickly, the response of the human body, such as the genes or epigenetic program, cannot keep up with it if we continue the current strategy. We need to find a way to enable adaptation of the human body program to change. Medical education should also be changed to focus on the harmony and balance of complexly entangled systems. Addressing such a need for changes, "systems medicine" is a new approach to achieving breakthroughs in disease prevention and treatment by obtaining and using fairly comprehensive information, such as molecular-, gene-, and cell-level information, as well as lifestyle, clinical, physiological, and environmental information.

From disease-centered strategy to comprehensive health management!

Looking back at the history of medicine, it was not far in the past that the treatment of disease based on accurate diagnosis emerged as the pivotal part of medical practice. Humankind was able to enjoy the fruits of scientific achievements in earnest beginning in the 19th century, with the discovery of pathogens, the understanding of cells and tissues, and the invention of medical devices such as sphygmomanometers, microscopes, and radiographic devices, which prompted innovations in diagnosis. Antibiotics for infectious diseases and therapies for the treatment of chronic diseases were developed in the 20th century. Therefore, it was not until the 20th century that the scientific way of diagnosis and treatment of diseases was firmly established as an essential part of medicine.

The characteristics of disease changed as well with the arrival of the age of chronic diseases after the age of infectious diseases. Today, we observe a spectrum of diverse diseases that cannot be properly managed with a simple strategy developed in the age of infectious diseases. In fact, it is common for a disease to appear along a continuous spectrum of clinical manifestation

The Changing Era of Diseases. https://doi.org/10.1016/B978-0-12-816439-6.00004-1

even if it is classified as belonging to a specific disease. For example, elevated blood pressure is usually diagnosed as hypertension when the systolic blood pressure is greater than 140 mm Hg and the diastolic blood pressure is greater than 90 mm Hg. If the systolic blood pressure is 130 mm Hg, however, and the diastolic blood pressure is 85 mm Hg, it is difficult to consider it a nonhypertensive state because it has been shown that people with a blood pressure of 130/85 mm Hg are more likely to have a shorter lifespan and are exposed to higher chances of suffering from various other chronic diseases, such as heart disease or stroke, than those with a blood pressure of 120/80 mm Hg, which is deemed the normal systolic/diastolic blood pressure. Therefore, the dichotomous strategy used to diagnose disease or consider treatment options may be just one way of efficiently managing disease but is not necessarily a sufficient strategy to properly manage a person's health. In that respect, it is not enough to just dichotomously divide a person's medical condition into hypertension and nonhypertension; rather, one has to understand the person's lifestyle, environmental factors, and genetic factors affecting blood pressure along a continuous spectrum and to eliminate all the factors that may raise the person's blood pressure.

It was only in the 19th century that diagnostic names of disease like "hypertension," "diabetes mellitus," and "heart disease" began to be used in earnest. Before that time, medical terms based on phenomena or symptoms that can be observed only with the eyes, such as diarrhea, weakness, fever, and edema, had been used extensively. In fact, the use of modern diagnostic names like "hypertension" was aimed at ensuring efficiency in disease treatment by classifying diseases with the same pattern or clinical findings as belonging to the same disease group. In other words, the aim of this approach was to improve the efficiency of treatment by applying a standard treatment to the same group of patients. This efficiency, however, does not necessarily improve the level of medical care for individual patients because it is very common for patients with the same category of disease to demonstrate different disease expression.

Contracting multiple diseases at the same time is also common. A deeper look at the disease called "hypertension" shows that it is often accompanied by diabetes mellitus and heart disease. In particular, as the elderly population increases, the number of people with multiple chronic diseases is increasing as well, making it impractical to cope with patients' problems using single-disease–centered strategies alone, which focus on specific body systems, such as circulatory or nervous system, or body organs, such as the heart and kidneys. In addition, patients do not just want treatments that help relieve them of the various symptoms stemming from their multiple diseases; they also want to be treated conveniently and effectively. Therefore, diagnosis of patient disease should be made by reviewing the individual situation of the patient in a comprehensive manner rather than concentrating only on the disease manifestation, while "comprehensive health management," including methods for overcoming physical, psychological, and social dysfunction, should be performed as well.[83]

Therefore, it is necessary to comprehensively identify and manage various lifestyle habits and environmental factors that cause disease or exert an influence on the disease progress. In other words, disease diagnosis and treatment, which are at the heart of today's clinical practice, should be seen just as a part of comprehensive health care. Probably, future health care will include the management of higher-level or more complicated issues, such as determination of longevity, surgery and prescription for maintenance or enhancement of human function levels, and management of the death process, as well as the conventional preventive medical services, such as lifestyle recommendations, regular health checkups, nutritional prescriptions, and genetic testing. With these technological advances and the changes in health care, humankind will greet an age of disease control and longevity determination with practices very different from those in the past.

The human body program must adapt to the changing environment

Because our living environment changes so quickly, the response of the human body, such as the genes or epigenetic program, cannot keep up with it if we continue the current strategy. The few available options left to us are to slow down the speed of environmental change or to enable adaptation of the human body program to the change. Evolutionary history has shown that the individual organisms or the species that succeed in adapting to the changes will thrive while enjoying the changed environment, but those that fail to adapt to the changes will become extinct. In Lewis Carroll's novel *Alice Through The Looking Glass*, the Red Queen advises Alice, "You must run as fast as you can, just to stay in place."[84] As everything is changing constantly, we cannot stay in place because we will lag behind the others if we try to stay in the same place because human beings are not independent beings but are just parts of the whole system. As the whole system changes over time, staying in the same place without corresponding effort to change means that it will not be able to play a given role within the system and will soon disappear.

A chronic illness occurs when the environment surrounding humans and their body program fails to maintain the balance between them, impeding the normal functioning of the human body program. Ever since the advent of civilization, especially since the Industrial Revolution, the pace of environmental change has been so fast that the human body program pegged on genes was not able to adapt to the rapidly changing environment, prompting the occurrence of chronic illnesses. If such changes continuously occur at a rapid rate in the future while the human body programs have yet to adapt well to such a changing environment, the chronic disease resulting from genetic and environmental incompatibility will never be overcome. In this case, humanity will continue to suffer, bound by the inescapable shackles of disease. At present, the only way out of this trap is by adapting to the changing environment, which means that the human body programs must be changed accordingly. Considering the current advancement of science and technology, humankind ultimately may end up succeeding in changing the human body program to escape chronic diseases. Otherwise, humans will continue to suffer from chronic illnesses because of the failure to adapt to the change. Therefore, humanity must find a way to make the human body program, including the genes, adapt to the changing environment to end chronic diseases.

The method, however, is not so simple. Basically, there are two methods: (1) correction of the human body program itself by using medicines, surgery, or genetic manipulation and (2) continuously changing each person's life and behavioral style to suit the changed environment, using the various pieces of information being churned out relentlessly. The method of making the human body programs adapt to the environment, however, will have difficulty to succeed if we take a simple concept that correction of specific factor is the solution for a given problem. For instance, attempting to prevent or treat Alzheimer disease by genetically manipulating the mutated part of the *ApoE* gene, which is related to the development of Alzheimer disease, turning it into a normal gene, does not necessarily mean that Alzheimer disease can be cured, the reasons being that each of the systems affecting the Alzheimer disease consists of very complex programs beyond the *APoE* gene mutation and that complex systems are again linked with one another in a parallel or hierarchical manner, interacting with each other and affecting overall health status. Therefore, for the complex system to operate normally in harmony and balance, there should be in place a more comprehensive and sustainable environmental adaptation program.

The first step for making the program successfully in motion is to provide customized information individually to every member of society. In other words, it is necessary to provide information individually about the complex disease factors and the ways to manage them precisely. However, it is not uncommon today to find cases where the information about the living environment that affects people's health is not adequately handled, sometimes causing confusion, and is not leading to meaningful behavioral changes that can effectively prevent disease and promote health. For instance, let's look at the information about cancer. Information on many food, such as broccoli, tomatoes, green vegetables, and yogurt, and lifestyle habits that are known to prevent cancer is delivered to the public almost daily via the mass media, along with the information on numerous carcinogenic factors, including cigarette smoking, alcohol drinking, chemicals, and ultraviolet light. A variety of information on cancer treatment is also being poured out without being screened. It is hard to say, however, that such information is actually used to prevent disease and promote health, since many of the factors are entangled with each other and may affect each person differently. Therefore, this information, to be truly effective, should now be tailored to individual health conditions.

Need for urgent changes in medical education

If we listen closely to the argument of the American physicist and historian Thomas Kuhn that the scientific paradigm is open to competition with different ideas and rationales, it is evident that scientific methods and thoughts do not exist as undisputable and absolute truths but as chains of processes in which multiple ideas compete with one another before one that best fits the given social context or historical environment is selected.[85] Now, medical education should also come out of the framework of classic biomedical models based on the simple causation theory that a specific factor causes a specific disease and the pathologic locus theory of diseases, in which a specific disease appears in a specific organ. Medical education itself has to change in line with the paradigm shift to "systems medicine," which focuses on the harmony and balance of complexly entangled systems, where various factors are networked with one another and affect various diseases.

After the publication of the Flexner Report in 1910, which emphasized that "medicine" should be considered a science rather than a mere clinical therapeutic discipline, medical education has been implemented in a form that combined basic medical science, which explores the root causes of diseases, and clinical practice, which treats diseases in the hospital, in the vast majority of developed countries in North America and Europe.[86] This system has continued to this day, exerting a profound impact on the current medical education. In fact, it cannot be denied that biomedical model and the medical education based on the Flexner Report have made a significant contribution to the development of modern medicine. In particular, it has had remarkable achievements in containing infectious diseases, thereby ending the age of infectious disease epidemics. In addition, the development of numerous drugs and medical technologies has provided a background for considerable success in the management of noninfectious diseases, such as chronic diseases. Due to this achievement, the incidence of infectious diseases has declined since the mid-20th century, but such development has not yet prevented a dramatic increase in the incidence of chronic diseases like diabetes mellitus, heart disease, and cancer. In other words, humanity has not reached the goal of preventing the occurrence of chronic diseases or controlling them completely to help restore patients to a disease-free and healthy condition. Therefore, acknowledging the limitation inherent in the current model of disease causation and medical education, we need to switch it from the medical education based on the biomedical model to one based on the systems medicine concepts.

Today, medical students learn myocardial infarction by understanding the chest pain and related symptoms accompanying myocardial infarction as well as the other symptoms that they need to distinguish from it. They are also taught diagnostic methods such as electrocardiography, ultrasonography, and cardiovascular imaging, as well as drugs and vascular stenting. They are not likely taught, however, about the factors affecting the cardiovascular status, such as the environmental factors; the constitution of and changes in the genes, epigenetic programs, and metabolic profiles; inflammation and toxic metabolism; mitochondria and energy metabolism; and the relationship of the cardiovascular status with other diseases. Moreover, education on the psychological and social problems experienced by patients suffering from myocardial infarction and the role of the physician as a patient-care, not just as a disease-treating, specialist is hard to find. Of course, if we take this approach, the volume of education is too vast to be realized for each and every disease. In fact, the notion that everything corresponding to a specific disease should be taught is no longer valid because diseases are not completely independent of one another and many things shared by different diseases will be unnecessarily repeated in their teaching. Therefore, the future medical education should be made different from current one so that students will learn each system inside and outside the human body, understand how these systems affect one another from the beginning to the eventual occurrence of illness, and learn the complete management of diseases, from disease prevention to treatment, all centered on the patients, not individual diseases.

In addition to medical education, hospitals must change as well. Since the 19th century, hospitals have become facilities for performing clinical tests and associated treatments based on them, rather than simply taking care of patients. Hospitals have now settled down as institutions where various tools and equipment are used by the professional medical staff to diagnose and treat diseases. Meanwhile, in hospitals, disease-centered care has been widely performed, and the patient is regarded simply as a person with diseases to be treated. While we cannot deny that the disease-centered treatment strategy has achieved considerable success, it has also revealed quite a number of problems. Currently, more than half of the elderly have chronic diseases, with many suffering from two or more chronic diseases at the same time. As a result, people with multiple illnesses have ended up being treated for each of their diseases separately. In this case, the disease-centered treatment of people with multiple illnesses not only is inefficient but also can cause confusion. This inefficiency and confusion will become more prominent during the transition period from the age of chronic diseases to the age of late chronic diseases.

In the end, hospitals should convert their clinical practices to one that focuses on treating patients rather than on individual diseases, and the way to educate and train physicians should be changed accordingly. When physicians do not understand the whole intertwined system but deal only with a specific disease, it is like walking through a new forest looking only at the trees therein to find a way, without seeing the entire picture of the forest. Probably, they are easily lost on the way! The plan to educate them to understand each disease as an independent unit and to train them to become experts in the specific field, as in the current physician training system, may have succeeded in producing disease-specific specialists to a certain extent. However, it cannot be deemed to have been successful in producing well-rounded experts who will greet and treat patients in the age of late chronic diseases because the training program for physicians does not fit the paradigm required by the changes of the times. Future physicians should understand the importance of maintaining the balance and harmony between the internal system and the external environment and the various phenomena that occur when such balance and harmony are broken and should be taught the knowledge and skills that will enable them to adopt an integrated and holistic approach when managing their patients.

Future hospitals and medical systems

There are currently many occurrences of misdiagnosis or inappropriate medical practices being carried out even in the modern hospitals and many patients experience deterioration of illness and even death there. Why are these so common in the field of modern medicine, which introduced medicine as a science and the disease-oriented clinical practice as the core approach? The reason for this is that medical practitioners often do not carefully observe the patient's symptoms and test results and fail to properly assess the patient's circumstances. On the other hand, that is also because it is not easy to equip the capacity to enable accurate decisions about patients in a short time period of their training, although the time is relatively longer than the training period for other disciplines. In fact, the amount of information that a physician needs to know is so great that it is almost impossible to have the ability to know it all. It is also extremely difficult to acquire the new information being discovered every day and to use it for diagnosis and treatment properly. On top of this, even well-trained physicians cannot escape from errors that any human being can commit during medical tests, prescriptions, or surgery, which often result in serious damage to the patients.

On account of all these problems, we will soon see the day when computers replace a significant portion of physicians' medical practices, such as diagnosis, prescription, and surgery. This does not necessarily mean, though, that the role of the physician will be diminished. Although the technical part of the medical practice will be performed by computers and robots, physicians will act as the overall health care providers for the patients. In fact, physicians can diagnose and treat their patients more accurately and effectively by using computers and robots as assistants. Using a program that supports diagnosis and prescription decisions based on precise information about the patients, physicians will be able to treat each patient more appropriately according to his/her characteristics and environment. Watson, a supercomputer-based artificial intelligence (AI) already developed by IBM, has the ability to gather vast amounts of medical data to provide the medical evidence needed for physicians to make a decision, and in some areas, it is capable of providing more accurate diagnosis and treatment decision than veteran physicians specializing in the corresponding field. Therefore, if this kind of physician-assisted program is used well, physicians will be able to spend time more taking care of their patients because it will take much less time for them to obtain information from their patients and to analyze and judge it. In a word, it would be the technical foundation for the transition from disease-centered care to patient-centered care.

If this trend continues, the hospital itself will turn into a huge automation system. Nearly all of the processes for diagnoses and prescriptions will be automated, and computers and robots will perform most of the tests and operations. In addition, various diagnostic devices will be equipped with algorithms that can make a judgment, beyond current role of merely presenting their findings. For example, if a patient undergoes magnetic resonance imaging or an ultrasound scan, the diagnostic device will help the physician make a diagnosis by presenting diagnostic names, way beyond merely showing pictures or results on the screen. Along with the prescription, various instructions for a desirable daily life will be automatically entered into the patient's information terminal, and the necessary medications will be automatically transferred to the patient's home. In the case of surgery, it is not necessary for the surgeon to manipulate the robot, as in today's robotic surgery; the robot will be so advanced that it will be able to make a judgment by itself and proceed with the operation. In this scenario, the medical staff will play the role of guiding the patient appropriately and managing the entire operation process to ensure that the preoperative treatment, the process of the operation, and the postoperative care are carried out as planned.

The continuity of care can be ensured even when the patient is transferred from a local clinic to a hospital or a general hospital in case his/her illness is difficult to treat in the local clinic. Suppose that a patient who has received continuous monitoring and management by the medical staff in a community clinic needs to be transferred to a higher medical institution because it has become necessary for him/her to go through a complete medical checkup or difficult surgery. In this case, the local medical staff will use the facilities and equipment of the hospital directly or will treat the patient in cooperation with the hospital staffs. The information about patient care is shared by the entire medical staffs through the computer network system, and this information is used to make final decisions, thereby enabling accurate, consistent, and comprehensive treatment for the patient.

As such, comprehensive medical care can be thought of as a system in which the health management continues from home to the hospital. To achieve this, people from different fields of expertise, from home to the school or workplace, and ultimately those who work at the hospital's intensive care unit, should work together. Therefore, the patient information should be transmitted from home to the hospital, and the information sharing and judgment system should be programmed in such a way that would facilitate cooperation among them to make the appropriate judgment for the patient. This is another technical basis of continuous and comprehensive medical care. Of course, measures to protect the patient's personal information throughout the course of the treatment will also be secured through technological advances.

Patients should also continuously transmit their health information to the medical staff via a monitoring device implanted in their body or a biomaterial analyzer placed in the toilet, as well as via wearable mobile transmission devices, such as clothes, watches, and glasses. If any abnormal finding is noticed from such information, a readily available medical treatment system will be provided immediately. The health status of each patient will be judged in connection with the health information that has been monitored since the patient was a fetus in the womb, along with the patient's genetic and living environment information and the personalized health management guidelines. In the future, the data on the patients' genes, living environment, past diseases, germ types and distributions in their body, and chemical substances they are exposed to as well as data from the biomaterial analyzer at home will be integrated. The integrated and properly interpreted data will be provided as reference data for medical judgment, thereby eliminating misdiagnosis or inappropriate medical practices. As such, physicians will be able to provide medical services in earnest centered on patients rather than on diseases.

A shortcut to ending disease: systems medicine

Chronic or late chronic diseases are caused by complex exposure factors interacting with the genes, which prompt a reaction in the body beyond the normal range of action, thereby negatively affecting the body structure and function. Care based on the principle of systems medicine can be called "personalized medical care" in that disease prevention and treatment are provided to an individual considering all the disease-related factors of the person. In fact, personalized medicine is not entirely new. The blood transfusion process, through which the blood type of the patient is determined and then the blood that matches the identified blood type is transfused to the patient, can be deemed as a form of personalized medical care. If blood type A is given to a person who has blood type B, a severe immune response may cause the death of the patient. Therefore, the patient should be able to receive blood that is appropriate for him/her, instead of receiving any blood. This is a somewhat extreme case, but if you do not know the patient information properly, you will not be able to achieve a desirable therapeutic effect but may end up causing a serious side effect, which again can worsen the patient's health even further. The same is true for the efforts to prevent disease. If you give standard precautions without knowing the patient's precise information, you can worsen the person's health even further in some cases. As many factors affect a person's health, it could be dangerous to consider only certain factors and provide preventive guidance according to the limited information.

Systems medicine is a new approach to achieving breakthroughs in disease prevention and treatment by obtaining and using fairly comprehensive information, such as molecular-, gene-, and cell-level information, as well as lifestyle, clinical, physiological, and environmental information. This approach considers a comprehensive array of information that can affect each person's health instead of the approach based on the concept of a simple relationship between specific factors and specific diseases, as outlined in Chapter 2. Therefore, for systems medicine to be realized, the information on genes, metabolites, and biomarkers should be considered together with a wide variety of exposure information, and the information management basis for processing such complex data should be well established. In addition, advanced data processing skills are required because large amounts of data ("big data"), including lifestyle, environmental, biological, microbiological, and clinical data, should be analyzed as combined or merged data.

This approach may deal with information significantly different both quantitatively and qualitatively from the information currently available for medical care. For example, the detailed analysis of blood cell types and their distributions may be used to measure a person's immune status instead of simply measuring the number of blood cells. Blood pressure, serum glucose level, and heart rate information may be automatically transmitted through a portable device and then monitored continuously. In addition, it will soon be possible to know what kinds of exposure factors or diseases a specific person is vulnerable to by analyzing his/her genetic and epigenetic profiles. It will also be possible to identify the patient's potential risk for certain diseases by evaluating the microbial patterns in the patient's colon through the analysis of his/her feces. Blood analysis can be used to diagnose cancer at a very early time by detecting abnormal gene fragments circulating in the blood. The various clinical tests that are currently being used for health screening will be further expanded and refined to provide personalized information on almost every possible disease of the person on a regular basis. As such, the precise knowledge on individual's current health status and the detailed information about the person's exposure will make customized health care realized.

There is a condition, however, that should first be arranged for systems medicine to be properly realized: the information on the determinants of diseases, which can be acquired through detailed evaluation of network of complex systems of the inside and outside of human body. The reason for this is that it is only after sufficient information about the pathogenetic mechanism, from the disease initiation factors to the reactions in the body and to the eventual disease occurrence, has been accumulated that one can determine which part of a person's body or which lifestyle or environmental factors to be corrected for the prevention or treatment of the disease. Therefore, efforts should be made a priori to obtain proper information for prevention and health management appropriate to individual characteristics through systems medicine research.

To see how the various biological reactions in the human body, especially those stemming from the exposure to the lifestyle and environmental factors, change over time alongside the life stage, we must track large populations with different characteristics over a long period of time and analyze the results. Therefore, large-scale and comprehensive follow-up studies throughout various age groups, from fetuses to elderly people, are required to track and investigate population groups with different racial and cultural backgrounds. The level of prevention and treatment of chronic diseases will be dramatically improved if the information obtained from such studies can be quickly shared and used by medical professionals around the world. The systems medicine based on this global medical information–sharing system, and the precision medical approach based on this system, will enable us to control chronic disease in the near future.

4.2 Global Strategy for the Borderless Disease

Community needs to take a responsibility to secure an environment in which each individual can prevent disease and promote health. As communities interconnected without physical barrier become the foundation of the future society, medical institutions will also be interconnected each other, thereby forming a comprehensive medical system. It is likely that the city will become a unit of disease management system in the coming world. If cities are tied closely together, however, along with globalization, this can become another risk factor for disease. Just as urbanization provided a hotbed of chronic illnesses, globalization may be a trigger for new epidemics of infectious and environmental diseases. Therefore, the next step is to create a strategy to end the disease on a global scale, based on a strategy of the globalization of health care to reduce the difference of health and disease levels among countries.

Personal practices and community-level efforts to end disease

The disease management strategy that is most urgently required now for ending the era of diseases is to have a system in place for appropriately managing chronic or late chronic diseases, which currently bring the greatest distress to people, with their increasing prevalence. The management of these diseases begins with providing all the information that the person needs and helping the person make appropriate decisions and take necessary actions based on such information. To recap, it is necessary to have a system that continuously monitors the physical, chemical, and biological factors of the surrounding environment as well as the internal body of each person and to help him/her make a judgment based on the information and act accordingly.

Let's imagine a man walking or running in the park. If a system is developed that closely monitors the information on the animals and plants in the park as well as the information on the microorganisms therein, and that analyzes the potential factors that can pose a threat to humans, and if such information will be provided to the man, then he will be able to judge what to do next and act appropriately for his health. The system can also monitor all the chemical substances that people can be exposed to, as well as air pollutants like fine dusts and ozone, before giving a warning when the level of a specific component in the air is high. The system will also tell the person about the expected weather condition, such as snow, rain, temperature, and humidity, as well as the corresponding measures to take to maintain the optimal conditions for his body. Based on such information, the person will decide how long his walk or running will be, which way to go, and when to rest, by taking into account his genes, defense capacity to toxicity, and health conditions.

The continuous reporting of the monitoring results, however, including the information on the environment outside the human body, such as the air pollutants, toxic chemicals, and weather factors, as well as the information inside the body, such as the genes, epigenetic profiles, metabolites, inflammatory status, and oxidative stress, does not guarantee the prevention or management of chronic or late chronic diseases automatically; this will not be possible unless each person puts into practice measures for improving his/her lifestyle and environment factors. In addition, as the number of elderly people is increasing along with the increasing human lifespan, chronic or late chronic diseases are more likely to occur. Therefore, the incidence of these diseases can be reduced only if people exert more efforts than before. It is, therefore, important to make people understand the living environment factors associated with the occurrence of these diseases and exert efforts to address such factors.

Let's first look at food. The goal of long-term dietary management should be to reduce the risk of acquiring the chronic or late chronic diseases that the dietary habits of modern humans can cause. Therefore, it is necessary to formulate desirable dietary habits based on a full understanding of humankind's dietary habits, which changed from the one in the age of hunter-gatherers to a grain-oriented one after the Agricultural Revolution, and to another characterized by a markedly increased intake of animal fat after the Industrial Revolution, which had a considerable effect on disease development. Also, it is necessary to continuously do regular exercise, such as running, brisk walking, and the lifting of proper weights, to maintain a healthy body because our genes require a certain level of physical exercise. In addition, although it is virtually impossible to eliminate stress, which is one of the key factors for the development of chronic or late chronic diseases, it is necessary to manage stress appropriately so as to maintain a normal psychological state: combination of relaxation of tension with intermittent tension.

The living environmental factors that cause chronic or late chronic diseases, however, are not limited to individual lifestyles. Factors like climate change, environmental pollution, and the increase of chemicals in the vicinity of our residential or working place also significantly contribute to disease occurrence. These factors are difficult to address with individual efforts alone. It is therefore necessary for the community to take a responsibility to secure an environment in which each individual can prevent disease and promote health. The environmental chemicals will continue to increase, and some of them may be toxic or carcinogenic or may even interfere with the normal hormonal actions. Moreover, this may affect not only the health of a particular individual but also the health of an entire population in the community. In the end, the community should play a role in reducing chronic diseases by providing a safe environment without necessarily creating inconvenience in daily living, through the stringent management of the production, distribution, and use of environmental chemicals. For this, the community is required to exert considerable efforts to reduce hazards of the chemicals for promoting the health of its members, because, in some ways, the environmental pollution or chemicals created by the community have a greater impact on the community members' health than individual choice of their lifestyle.

Urban environment determines our health

Humankind underwent major changes in the form of community until majority of communities were eventually formed in cities. The Agricultural Revolution that prompted the formation of civilization, and the Industrial Revolution, which led to modernity beyond the Middle Ages, changed the shape of the human residence twice. The first change was the shift from the life in the age of hunter-gatherers, where people could not settle down in a certain place for a long period, to permanently settled living in a specific area along with the agricultural revolution. The second change was the rapid increase in the urban population as a result of the massive migration of people from the rural areas to the cities along with the Industrial Revolution. As a result of these changes, humankind's residential style changed dramatically from the family-centered residences in the age of hunter-gatherers, where one or a couple of families lived together, to a village-centered residential style in the wake of the agricultural revolution, and to the urban residential style characterized by a rapid increase in the population per unit area and the clustering of strangers.

Urbanization, which has expanded even more since the 19th century, has had a considerable impact on the health and disease patterns of modern humans. In addition, to the introduction and spreading of new pathogens due to the increasingly close contact among people, the city provided sufficient risk factors conducive for the occurrence of diseases, such as a decrease in physical activity caused by convenience facilities and an increase in air pollution caused by toxic gases and dusts emitted from factories or automobiles, all of which ended up becoming critical factors that decisively increased humankind's chronic diseases. In fact, the diseases that are epidemic in the modern society, such as diabetes mellitus, hypertension, cardiovascular disease, cancer, allergy, and depression, are not due so much to the changes in individual lifestyles as to the changes in the residential and living environments of the community, such as urbanization.

Nowadays the city has become the central form of community. Probably, it is more reasonable to think that urbanization has not progressed steadily since the beginning of civilization but has progressed rather rapidly of late after stagnation for a long time. Three thousand years ago, there were only four cities in the world, with their respective populations greater than 50,000, and by the year 2000 BC, still only about 40 cities had a population greater than 50,000 people. Population migrations and exchanges exploded during the period from the Age of Exploration until the imperialist era, and many people began to live in cities, but the urban population still remained at only 5% of the world's population until the beginning of the 19th century. Then urbanization progressed at a rapid speed in the wake of the arrival of the modern society after the 19th century on, and within just 200 years since that time, the urban population has grown remarkably, accounting for greater than 50% of the world population at the moment.[87] Such rapid urbanization is the main reason for the incongruence between the genes and the environment to transpire. Therefore, what is more important than individual efforts to change lifestyles are the efforts to reduce the incongruence between human genes and the modern environment by building healthy community from the early stage of urban planning and community development.

With the progress of urbanization, the sizes of cities have grown enormously, and dozens of large cities, each with a population of greater than 10 million, have appeared all over the world. Although these metropolitan changes seem to be prominent now, however, this trend is "unlikely" to continue for a long time. In fact, the need for a large and physically congregated city is likely to decrease in the future because as even physically distant places are interconnected with one another via computer networks or very fast transportation, there will no longer be any inconvenience in life in such places. The quality of life can be further improved as there will be no restrictions imposed by physical distance in accessing medical services as well as administrative services and education. In the end, the urbanization trend in the future society will gradually shift from urbanization centered on big cities to one centered on small communities interconnected each other.

As communities interconnected without physical barrier become the foundation of the future society, medical institutions like clinics and hospitals will also be established interconnecting each other, thereby forming a continuous and comprehensive medical system responsible for the health of the community members. As hospitals are automated further, computers and robots will perform main works for diagnoses, prescriptions, and surgeries. The physicians working in community clinics will be transformed into those responsible for managing the overall health care, as family physicians. Perhaps the future medical institutions will no longer be the place for patients to visit after judging their own health problems upon the appearance of symptoms, as it is today. Instead, the medical institutions like clinics or hospitals will monitor the environmental factors, lifestyle, and biomarkers of each individual using the community's computer network system, and if any abnormal signal is detected, it becomes the place where the problem will be solved by guiding the patient even before he/she feels the symptom. In addition, the community environment, such as housing, transportation, food supply, waste treatment, green space, and clean air and water, will be planned and implemented centered on the community members' health.

This change, however, will not happen by itself. Desirable changes will occur only if planning and implementation efforts is made to construct the network systems where the elements constituting the foundation of the community will be interconnected centering on the community members' health. In other words, efforts should be made to create an urban community environment with human health as the overriding value, and based on that, a comprehensive medical system connected by a computer network of clinics and hospitals should be created.

A well-organized community network in the future will be based on a dramatic increase in productivity, and the remarkable development of science and technology will lead to new changes in the people's way of life as well as in the future disease management. It is difficult to predict future changes in detail, but the existing institutions, such as the state, church, and the social status, will collapse or at least lose their power, and a new order will rise to replace them. In fact, it can be said that the existing order was intended to maintain the production relations corresponding to the productivity level. For example, the concept of the state came about in the modern society because the capitalists wanted to protect their interests through the state. As it is no longer possible to maximize the interest of capital within the framework of the state, however, people pushed for the globalization of capital, and the state is gradually losing its power. This implies that the basis of productivity has been changing from a framework based on the state to an expanded framework beyond the states, and therefore the production relation— i.e., the relationship between capital and labor — is also being globalized.

Therefore, while the presently powerful state framework is getting weaker, the urbanization will progress even further, and the urban community will develop into a more independent form, thereby becoming a basic unit of globalization. In the end, the vast majority of residential areas in the world will be urbanized, and the city will become a unit of disease management system as well as an economic, political, and cultural unit in the coming world. If cities are tied closely together, however, along with globalization, this can become another risk factor for disease.

The danger of globalization: the era of borderless disease

The modern society built on industrialization and urbanization has led to an epidemic of chronic diseases along with the remarkable achievement of a dramatically increased lifespan. The modern society also stepped into the era of globalization, without stopping at industrialization in each country, where the world is bound to one market, production is outsourced to other countries, and sales are made to global consumers. Globalization is the process by which the entire world is interwoven as a single connected community as trade and exchanges grow beyond the level in the past. Just as urbanization provided a hotbed of chronic illnesses, globalization may be a trigger for new epidemics of infectious and environmental diseases.

For example, patients with encephalitis accompanied by severe muscle weakness were reported at about the same time as dead crows were found in New York City in August 1999. The West Nile virus, first discovered west of the Nile River in Africa, entered the Americas when interregional exchanges became active, infecting people through birds. The West Nile virus further expanded its habitat in the Americas before spreading later to Europe as well as to Asia and Australia. The swine influenza that went pandemic in 2009 was the result of an infection that caused flu in humans by a virus that had taken pigs as hosts in the past. When the virus that originally inhabited in pigs were transferred to humans, mutations happened to occur in the virus which makes the mutated virus transmissible among humans, and the frequent travel between countries or regions led to a pandemic spread of the flu.

The West Nile virus and the swine influenza are cases where viruses from animals have caused epidemics in humans. During the 10,000 years of civilization, humankind has experienced a number of potent infectious diseases, most of which have been transferred from animals to humans when close contacts happened. Therefore, there are many more possibilities of such epidemics in the future because there are some areas that have yet to be developed, and there will be increasing contact with animals; further, human interactions will become more frequent along with globalization. There is also a great potential for new viruses and bacteria in the frozen ground to surface in the wake of global warming due to climate change and infect humans through insects and animals. With the predicted decline of the diversity of the ecosystem and animal species, the conditions of pathogens' habitats are expected to turn sour, disrupting the balance between pathogens and their hosts, thereby increasing the likelihood of the pathogens switching their hosts from animals to humans. Once humans become hosts of new pathogens, the likelihood of widespread epidemics as new communicable diseases will rise in the globalized contemporary society.

In November 2002, a disease characterized by fever and respiratory symptoms like pneumonia occurred in Guangdong province in China and soon spread to Hong Kong, Singapore, Vietnam, and then Canada. Although the host animal has not been clearly identified, it is presumed that the virus has migrated from a bat or a musk cat before causing the disease in humans. A total of 8000 people were infected, and among them, 800 died. The mortality rate seemed to be so high in the beginning that it was named severe acute respiratory syndrome (SARS). The Middle East respiratory syndrome (MERS), which became prevalent in South Korea in May 2015, was also caused by coronavirus that migrated from camels in the Middle East. It is now obvious that these new infections did not exist in the past, and the epidemic is no longer confined to a few countries. In fact, influenza, whose incidence is almost as frequent as that of the cold, has already become a borderless disease spreading to other areas as soon as it occurs in one area on the planet.

In 2008, a dioxin concentration greater than 100 times higher than the standard level was detected in pigs that were fed grains provided by Irish suppliers. As dioxin can cause cancer or diabetes mellitus, it could have a significant impact on the health of the Irish people. Moreover, Irish pork had already been exported to 23 countries by that time, making the problem spread beyond Ireland. In fact, fish and agricultural products, as well as meat, such as beef, pork, and poultry, are already becoming globalized, beyond the level of production and consumption within the local community or country. In addition, as the livestock feeds production industry supplies its products across borders or regional boundaries, any potentially harmful substance contained in the feeds can affect the meat produced in various regions, triggering a global spread of toxicity.

In April 1986, a nuclear power plant exploded in Chernobyl, Ukraine, a part of the erstwhile Soviet Union, releasing massive amounts of radioactive particles. The radioactive particles flew westward with the wind, polluting not only Belarus but also Russia and Europe. Thirty-one people died on the spot due to the explosion, but the number of cancer cases estimated to have occurred later due to radiation exposure was greater than 40,000.[88] The accident in the Fukushima Daiichi nuclear power plant in 2011 also showed this problem. The radioactive contamination was not contained in Japan; the contaminated water flowed into the Pacific Ocean, but we do not know what the consequences will be for now. Increasing amounts of air pollutants are also spreading widely across the borders. In South Korea, there have been frequent alarms on fine dusts of late due to the increased air pollution. This happens to a considerable extent because the fine dusts spewed from the neighboring industrial park in China flow eastward along with the westerly wind. These examples illustrate that a global community-level response beyond the boundaries of the region and country is required when it comes to responding to newly emerging diseases in today's world.

The World Health Organization declared in 1977 that it had eradicated smallpox, a move that was made possible largely by its extensive vaccination program since the mid-20th century. The vaccination program has contributed significantly to reducing the incidence of communicable diseases by preventing many diseases, such as measles, poliomyelitis, diphtheria, and hepatitis B. The use of antibiotics, such as penicillin, along with the success of vaccination, has also achieved great success in the treatment of infectious diseases, including communicable diseases. The rosy hope of ending infectious disease after conquering pathogens with antibiotics, however, turned gloomy again when humankind encountered an unexpected problem in the form of antimicrobial-resistant bacteria. In other words, most of the pathogens die due to the toxic environment created by antibiotics, but some genetic mutations occur in the pathogens with continuous use of antibiotics, resulting in the creation of bacteria that do not die in the toxic environment of antibiotics.

In fact, the antimicrobial-resistant bacteria that were created by genetic mutation can spread rapidly even if the number is small at first, because pathogens produce offspring very quickly. The other reason for the spread of the antimicrobial-resistant bacteria is that antibiotics are widely used in livestock breeding and fish farming, and are sometimes overused in the medical field. Such overuse of antibiotics has created favorable conditions for the bacterial mutations to occur more frequently. In addition, to pathogenic bacteria, pathogens such as viruses and malaria parasites are becoming increasingly resistant to respective antimicrobial drugs. As the past pathogens are reemerging newly armed with antimicrobial resistance, it may be difficult to achieve the goal of overcoming infectious diseases completely at least for the time being. Problems like antimicrobial resistance, however, are not limited to a particular country or region. In today's globalized world, it cannot be assumed that pathogens with an antimicrobial resistance can be contained within specific regions. Therefore, comprehensive global level strategies, beyond the local-level strategies, for suppressing the development of antimicrobial-resistant microorganisms are required, along with the development of new antibiotics and vaccines to effectively disarm the pathogenic germs.

Ending disease through the globalization of medical care

Disease patterns are basically determined by the development stage of civilization. However, due to the different historical development stages of each region and the different experiences and times of civilization construction and propagation, the patterns of disease may appear differently among various regions. At present, there are still family members or clan-centered societies living a life as hunter-gatherers, such as the Hiwi in South America, who have yet to suffer chronic diseases in earnest.[89] On the other hand, people in developed countries, while enjoying the modern urban living, are experiencing chronic diseases together with the late chronic diseases. These two populations, who are very different in terms of development stage of civilization, coexist. To cope with these different stages of social development and disease, strategies for dealing with diseases tailored for each society must be adopted.

If we look at the civilization of humankind as a whole, however, we can see that it has undergone specific stages of development along with the time. Even though there were some exceptional cases of having returned to hunting and scavenging when the condition was not conducive for farming and herding, such as in unfavorable weather conditions, there was no change in the basic direction of going forward to the modern society after the Agricultural Revolution and then the Industrial Revolution. Therefore, it is desirable to adopt a disease prevention and management strategy that is suitable for each development stage specifically because there is different disease experience at each stage.

A cross-sectional look shows that there are now various stages of disease on the planet, and therefore, it would be appropriate to adopt different strategies for each region according to the stage. It should also be taken into account, however, that the various stages of disease affect one another and that the disease stage changes along a certain direction. For instance, the problems of chronic disease in developed countries and nutritional deficiencies in sub-Saharan Africa coexist at the moment, and they may represent different stages of disease development, but the transition of disease pattern is not fundamentally different from one region to another because the sub-Saharan region will eventually experience an epidemic of chronic diseases soon, as in developed countries.

Therefore, despite the different stages of historical development among various regions, the trend of globalization requires a governance system capable of carrying out various strategies according to each stage in the global scale, and such governance system also requires close coordination between countries and regions. Along with the trend of globalization, the healthcare services that have been planned and performed mostly within the framework of the country need to be changed as well, embracing globalized approach genuinely. Otherwise, the world can be polarized more and end up with deepening inequality. In fact, many countries are now seeking to adopt the developed countries' advanced medical technologies while still struggling with poor medical access and a poor social environment. This may lead to the polarization of health care within the country, while accelerating the globalization of medical care at the same time.

Now, there is a need for global-scale disease management strategy that addresses these issues of polarization and globalization in earnest. Priority should be given to reducing the disparity in the health and disease levels in each country, leaving no one behind in terms of accessibility to medical care, by applying technologies and tools available to the local community. The next step is to create a strategy to end the disease on a global scale, based on a strategy of the globalization of health care to reduce the difference of health and disease levels among countries. Probably, one of the practical measures to conquer disease is to strengthen the international governance structures, such as the World Health Organization and the World Bank, that establish and enforce a global strategy.

4.3 Epidemic of Mental Illness Comes to Torment Humankind to the Very End

People's health status and lifestyle habits can be monitored via biosensors, and early measures can be taken to ensure that chronic diseases will not pose a threat anymore in the future. The changes in the coming years, however, will not simply bring about such a rosy future. Above all, the mental workload will grow excessively, although the physical workload will be reduced. This will pose a significant challenge because mental illnesses can increase significantly. As humankind becomes increasingly dependent on the technical development and becomes a part of a huge computerized network, people may feel threatened or anxious because of a loss about value of their existence as independent selves. Such change can lead to an epidemic of mental illnesses at a very alarming level. For this reason, humankind should prepare for mental illness, which will soon emerge as key disease in the hyperlinked society.

Disease management strategy in the wake of the network revolution

In 1981, *Time* did not choose a person of the year. Instead, it chose a machine of the year, signaling that the computer will be playing a central role in the future development of the society. Medicine, medical care, and disease treatments are no exception in that they are affected by this change. If the Industrial Revolution was a historical event that dramatically increased productivity supported by fossil fuels and machinery, the network revolution enabled by computers and the Internet is another major historical event that has brought about changes throughout all over the society. Humankind entered the age of abundance as the Industrial Revolution, which began in the 18th century, triggered a scientific revolution worthy of being called the "Second Industrial Revolution" at the end of the 19th century, but on the other hand, it caused the maladaptation of the human genes to the environment owing to the excessively fast-changing living environment. The decisive role that the maladaptation of the genes to the environment played in ushering in the arrival of the age of chronic diseases suggests much about the diseases that the network revolution, called the "Fourth Industrial Revolution," can bring about.

By the turn of the 21st century, the development of information technology had created a great turbulence in the society. With the arrival of the networked age, humanity began entering an era where people are connected with other people and with things and, where things are connected with other things, to enable a more intelligent judgment. Moreover, the speed of change is very fast, unlike in the past. It took 6 million years from the emergence of the hominids to the prehistoric humans to emerge, 10,000 years from the formation of civilization to the Industrial Revolution, and 250 years from the Industrial Revolution to the modern society. It took only about 30 years, however, for humanity to be interconnected with one another through the network revolution. The speed of change will be accelerated even more in the future until each person no longer exists as an individual independent from others but as a constituent of the global network that is one human community.

The characteristic of this age is that humankind manages the physical environment and the cyber environment in a consolidated manner, overcoming the constraints of time and space, and the entire planet is morphing into an organic system. In other words, the boundaries between people, objects, and events will be blurred, and there will emerge an increasingly close interconnection among them.[90] Therefore, people's individuality will disappear as everything is being integrated organically, while the system for managing this will also evolve from a centralized to a decentralized system of responsibilities. Ultimately, it can be said that humanity and the environmental factors will be linked with each other, transforming the society into a networked one that overcomes the constraints of time and space and creates new values.

In the networked society, all the information, such as how much people walk and exercise in a day, how much calories are consumed, how high the blood pressure is, how fast the heart rate is, and how many hours one sleeps, can be recorded in each person's smartphone or via IoT (Internet of Things) tools at work and at home. People's health status and lifestyle habits can also be monitored by detecting physiological or pathological changes via the biosensors installed in clothing, watches, and glasses, as well as via those inserted in the body or installed in the living space at home or at the workplace, and transmitting such information to the computer network. For example, the health condition can be continuously checked by inserting a small device capable of continuously measuring the blood sugar level or metabolites in the skin or by installing an analyzer capable of analyzing the DNA or microorganisms obtained from the urine or feces in the toilet. The monitoring via biosensors can be done automatically and inconspicuously, because it can deeply penetrate people's daily lives.

The information obtained from the biosensors will be automatically transmitted to the medical care system and analyzed so that if any anomalous signal is generated in the body, the medical staff that is in charge of the patient care will be provided with immediate information and the appropriate medical measures required for treating the patient. In fact, chronic diseases like hypertension, diabetes mellitus, and hyperlipidemia do not usually develop into serious problems if managed properly while monitoring the disease biomarkers accurately and continuously. Also, life-threatening diseases like cancer can be sufficiently cured if they are found at an early stage. Through such monitoring with biosensors, early measures can be taken to ensure that chronic diseases will not pose a threat to humans anymore in the networked future society.

Reduced physical activity and increased mental activity

In the future, the very shape of labor engaged in the production activity will change. The conventional practices, including going to work or attending a meeting, will almost disappear, and people will be connected to computers anytime and anywhere, working in a more efficient environment to gather information and opinions and then to make decisions based on these. The labor force will also change. The traditional way of working as a specialist in a specific field after mastering a certain specialty area will be gone. Instead, the task itself will change in ways that will integrate various fields. Therefore, opportunities for learning will be provided throughout one's lifelong period, even after the completion of a regular education course and in ways that will enable the acquisition of knowledge not just from one field but from various fields. Moreover, because elderly people will live longer in the future by strengthening their biological functions, age will no longer be a limiting factor to work, thereby gradually eliminating the concept of retirement.

If humankind's future productivity exceeds the level of human consumption and the community ends up having sufficient social infrastructure, it will be possible to provide all the members of the community with their basic necessities for daily living, including food, education, transportation, communication, and medical care. As the living standard of all the community members improves, the production relation will also undergo a similar transformation. In other words, humankind will be able to realize a community in which monopolization of ownership has been largely resolved. As the production and consumption are elaborately well managed based on sufficient productivity and advanced network system, the monopoly ownership over surplus products will disappear, along with the social class. If such a community is realized in a desirable direction, individuals can achieve self-realization through labor, along with a rediscovery of the true meaning of labor.

Therefore, if humankind successfully achieves such monumental transformation, the problems of slave labor, feudalism, and labor alienation, which have not yet been resolved through capitalism and socialism — i.e., "labor for others rather than for oneself" — can be essentially solved. Individuals can choose to work according to their abilities and aptitudes, and can change their labor activities as necessary. In addition, the work environment, which has been characterized by intensive labor and dangerous and repetitive work since the Industrial Revolution, will be turned into jobs that are less physical and taxing. With the introduction of remote manipulation and virtual reality (VR) technology, the industrial scene is being transformed into a smart space from a space where workers face machines directly. Thanks to the widespread use of state-of-the-art sensors and the interconnection between objects, plant operators and engineers will work in offices, just like white-collar workers. In addition, the office will no longer be a fixed place in the same company building, as it is today, but will change to flexible concepts like a mobile or home office.

The changes in the working environment, however, will not simply bring about such a rosy future. Above all, the workload itself may not be reduced because the mental workload will even grow in volume. As the amount of information that a person has to deal with is much larger than in the past, and as it is necessary to organize and analyze such information and to make decisions, the mental workload will grow excessively, although the physical workload will be reduced. This type of working environment will be markedly different from the labor condition in the age of hunter-gatherers, which was characterized by a massive amount of physical activity but a limited level of mental labor, which still defines our state of body and mind today.

In fact, it can be said that chronic diseases like obesity, diabetes mellitus, and heart disease are caused by the differences in the physical activity between the people in the age of hunter-gatherers and those after the Industrial Revolution. Likewise, if the physical activity is reduced further in the future while the mental activity grows in volume to an incomparable extent, this will pose a significant challenge to disease management because mental illnesses, which are often attributable to one's failure to accommodate the changes in the society or the relationship, can increase significantly. Accordingly, just as chronic diseases like hypertension and diabetes mellitus are caused by an imbalance between the energy supply and consumption with the considerable reduction of the amount of physical activity, the greatly increased amount of mental activity will cause an overload of brain activity and a corresponding explosion of mental illnesses like depression and adjustment disorder in the future. Perhaps the chronic disease that will dog humankind to the very end would be mental illness. Therefore, the future medical systems should adopt health management practices that sustain the human body and mind in a biologically optimal state by monitoring both physical and mental activities at the same time.

Increased mental labor shakes up the age-old physiological equilibrium of the human body

In the networked society, data and services can be used anytime and anywhere as necessary. As a result, the work efficiency can be significantly increased because the work is carried out without being limited by time and space, but the boundaries between the leisure and work hours can also become fuzzier.

You may have to work in your leisure time, thus tiring yourself with overwork. These increased tasks, however, are usually those that require mental activities instead of physical activities. The more computerization advances, the more the work style changes from physical activity to mental activity. This can shake up the physiological equilibrium of the human body that has evolved to the present form over a long period. That is, the amount of energy used by the brain has greatly increased compared with the past, while the amount of energy consumed by the muscles has dropped to a much lower level.

The muscles are very important in controlling the body temperature in addition, to enabling physical activity. If the external temperature detected in the hypothalamus of the brain (the central nervous system responsible for thermoregulation) is high, the blood vessels under the skin will be expanded to allow an easy escape of heat from the body. Conversely, when the temperature is low, it will shrink the blood vessels and reduce the possibility that the body heat is dissipated. In addition, heat is generated to maintain the body temperature by contraction of muscles thanks to the energy generated when the ATP is converted to ADP in the muscle. Therefore, shaking of the body in the cold is due to muscle contraction to generate heat. As such, the muscles play a very important role in controlling the body temperature. If the amount of muscle mass decreases along with falling energy consumption by muscle, the body temperature control function will deteriorate as well. As body temperature regulation is essential to maintaining the proper metabolism of the body and to facilitating key functions like the brain function, people will be increasingly dependent on external-temperature control devices like air conditioners, heaters, and clothes for controlling the body temperature. Of course, the devices for maintaining the appropriate body temperature can be further advanced as well. For instance, a sophisticated indoor environment controller or clothes capable of automatically adjusting the body temperature will be developed. This, in turn, means even more dependence on machines and tools for basic life-sustaining activities. Eventually, humankind will find it quite difficult to survive without machines or tools.

On the other hand, there is a limit to the information-processing capacity of the brain because brain size and the number of neurons cannot be increased. Therefore, the amount of workload that the brain has to carry out cannot continue to increase for good, suggesting that the amount of mental labor needed in the future societies will someday reach a point where the biological brain can no longer afford to handle it. In other words, there will come a day when a certain system capable of boosting the brain function will be needed. In addition, given the current pace of technological development, the brain's biological capabilities, such as memory, perception, and analysis, can no longer match the performance of AI. Therefore, we may wish to boost the biological capability of humans by using AI devices connected with the brain cells via interfaces. Perhaps it takes a considerable amount of time for AI to be equipped with deeper underlying biological capabilities such as emotions and moods to communicate with humans through such interfaces. Feelings and moods, however, are also derived from the interaction between the environment and individual humans as well as from the response patterns learned through biological reactions and experiences. Accordingly, AI will attain a certain level of capability someday that is good enough for it to understand and express emotions or moods.

Dependence on al lowers one's self-esteem

When the intelligence of our hominid ancestors and that of the modern hnmans are compared, a big difference will probably be seen, if not as great as the one between the apes and modern humans. The constant improvement of the intelligence is due to the pressure of natural selection, and the changes to further improve the intelligence will continue as long as there is competition for survival. Since the 1980s, the computer-enabled revolutionary changes have had a profound impact on everyday life as well as on the society as a whole, and a variety of attempts have been made to overcome the limitations of current biological intelligence. A look at our daily lives will show that we get information easily from our smartphones, store information on mobile media, and use the information stored in various media at any time, as necessary. People are now consciously or subconsciously exerting efforts to improve their intelligence so that they will not have to rely solely on their biological intelligence. These efforts are expected to accelerate further. In the end, it will reach the point at which AI will be used beyond the intelligence capabilities of biological humans.

In addition, studies to overcome the limits of life may lead to technological breakthroughs, such as that the brain cells will not die and will continue to function. If AI devices are connected with the brain via an interface, or are sufficiently miniaturized to enable them to be inserted in the brain, along with the use of technology to regenerate brain cells or to extend their lifespan, the human intelligence will be improved to a much higher level. If AI is combined with the human brain, the AI connected to the computer network will push the human intelligence to advance far beyond its biological bounds of individuals, and will further evolve to become the collective intelligence of humankind. This collective intelligence can create a new civilization residing in a world with completely different shape from the current civilization.

The aforementioned change may happen not in the distant future but in the present century. This, however, will not necessarily bring positive results only. Perhaps humankind will be biologically strengthened in the future, but at the same time, humankind will become more technology dependent and less different from machines. Also, as humankind becomes increasingly dependent on the technical sophistication and productivity of the community than on the individual's ability, the individual personality or independent self can lose its strength. For example, if the memory and intelligence will be strengthened by the AI devices implanted in the brain, the intelligence will be enhanced, but the independent self will not necessarily be strengthened. Rather, one's reliance on computers or machines can lower one's self-esteem and can make humans more of a component of the overall system because it will be difficult to play a social role without a computer, and the computer itself is not made by the independent individual but is a technical achievement of the community. Therefore, although the unity of the community grows, the awareness of the independent self, which has seen itself as a completed entity, will inevitably be reduced. In the end, people may hand over their individual independence to community as a whole and live just as components that make up the whole.

Just as a cell does not exist as an independent entity but only as a constituent of an individual organism where the cells unite with one another, an individual human also can be regarded to exist only as a social member and not as an independent individual. This is because each individual can have a sense of existence as a human being only when he/she establishes a relationship and plays his/her given role as a member of a family, as a close acquaintance with somebody, or as a part of a community. When a cell does not play its given role in the tissue, and has independence from the governing framework formed among the cells, it can become a cancerous cell that destroys the tissue, or it may turn into a diseased cell incapable of functioning normally, and then cause a disease or induce apoptosis at best, thereby killing itself. If a person does not play his/her given role in the community and crosses the boundary of cultural and social norms, he/she may harm the community or develop a mental disorder, such as an adjustment disorder, anxiety, and depression, or may even end up killing himself/herself.

In the end, for an individual to function properly between the two extremes of complete dependence and the rebellious independence, he/she should have a healthy level of self-esteem as an individual entity, and the relationships with others should be formed well harmoniously within the community at the same time. Therefore, the quality of life of the future humans can be determined by how much the individual freedoms are preserved and realized under the framework of the community or the networked society. It is not desirable for a society that individual freedoms are uninhibited or allowed to expand infinitely, but neither is it desirable for an individual to act only as a component of the community that loses its identity. Ultimately, maintaining an appropriate balance between human freedom and community control will turn out to be a key challenge in securing individual mental health and creating healthy communities in the future society. This may as well depend on how humankind manages AI, which will then have established itself as an important axis moving the community.

Existential anxiety triggers an explosion of mental illness

In 1950, Alan Turing proposed the Turing test as a criterion for judging whether a machine is truly intelligent. A machine that passes this test is considered to have an independent intelligence. After that, AI has continued to evolve with the development of computers. In 1997, IBM's chess-playing computer Deep Blue defeated Russian chess world champion Gary Kasparov with two wins, three draws, and one loss. In 2016, Google's Alpha Go defeated Korean Go champion Se-dol Lee in a historic match by winning in four of five games. As AI today is capable of performing complicated tasks that require a higher level of skills, AI machines will soon carry out a great deal of professional tasks that are currently being performed by humans.

The superconnected society that will materialize in the near future is a society where people act as nodes of a huge network, connected with countless objects and robots.[91] In this society, the meaning of existence as an individual human with an independent personality may become obscure, and the ability to live harmoniously with the superconnected system may be valued more highly. As humans have succeeded in building societies and creating complex relationships since the advent of civilization, they are expected to succeed as well in forming more complex relationships in the future, but moving toward a machine-connected relationship will pose a serious challenge to humanity. If you become a part of a huge network instead of existing as an independent human being and you feel threatened or anxious because of such a loss about value of your existence as independent self, it can lead to an epidemic of mental illnesses at a very alarming level.

The traditional powers appear to have gradually handed over their dominance on people to the networked systems. For instance, we can often see today that the social media like Facebook influences people more than a government power. It is not easy, however, to recognize such transition of powers, because it looks as though each individual seems to have taken on an independent position with free will. This is, however, in fact a shift from a subordination relationship to traditional power to dependence on the network system. The network can be regarded as a convergence of people from all over the world into one immense system. The main reason that traditional relationships like the state–citizens, employers–employees, and teachers–students have broken down and new relationships among people based on the networks have emerged is that we all have become subordinated to the complex connection of computerized networks. If an individual fails to secure his/her position in the network connection and loses his/her self-esteem amid the daunting feeling associated with the huge network of the superconnected society, he/she can suffer from an existential crisis, which easily turns into anxiety or depression.

In the hyperconnected society, an individual person does not make his or her own decisions about most issues, as the individual people today do; AI takes over it instead, making the vast majority of decisions, with the roles of humans being limited to the approval or understanding of such decisions. Humans can fall from the subjective position of thinking and acting to a passive and dependent position in biological and social terms. In short, humans with absolute and independent rationality (certainly it is a very hypothetical idea, but we have presumed as such!) may no longer exist, and there may only be passive and dependent entities subordinate to the gigantic network system. This threat will be different from any other threat that humanity has ever faced, such as hunger, germs, and chronic diseases, in the past.

Just as the changes in the period of 250 years since the Industrial Revolution have prompted a widespread epidemic of chronic diseases, the transition to a superconnecting society taking place in a very short period could bring a new disease epidemic in a large scale. That is, the insecure position of human existence can trigger an explosive outbreak of an epidemic of mental illnesses. Furthermore, based on the history of humankind, where collective psychiatric conditions often manifest themselves as serious political madness (see the Nazis and World War II), the widely prevalent mental illnesses in the future societies could have some catastrophic results. For this reason, humankind should prepare for mental illness, which will soon emerge as key disease in the hyperlinked society. Therefore, the future medical care should advance to the stage of managing the physical, mental, and social functions to fit the networked society, away from the present medical care limited to diagnosis and treatment of infectious or chronic diseases.

4.4 Economic and Social Inequities Lead to Biological Inequalities

If the rate of change in human society is too fast, it can cause serious problems leading to the crisis of human sustainability, way beyond health problems. If we address them successfully, we will be able to carve out an ideal future. For instance, humans can survive well harmoniously beyond the limits of their biological lifetimes. However, unless special efforts are exerted, there is little chance that the future will evolve as such. The more likely scenario, in fact, is that the productivity will improve further while worsening the overall social and biological inequality. The inequality may manifest as the difference between the biologically strengthened new human beings and the *Homo sapiens* without such strengthening. If we want to avoid such tragedy, we must turn the direction of change towards ensuring the sustainability of humankind.

Uncertainty of the future leads to humanity's crisis

With the exponential advances in science and technology, it is increasingly possible to treat chronic and late chronic diseases successfully and to overcome the limitations of life expectancy. Disease will not be terminated, however, and a happy future will not come so easily, even if the disease treatment technology will be improved to a level that can cure most of the chronic or late chronic diseases. The reason for this is that if the present economic inequality, the unbalanced development of science and technology, and the difference in medical accessibility are sustained or accelerated, there will certainly be population groups that cannot enjoy the benefits of the medical technology development. Although humankind has evolved into a global community, the conflicts between countries, religions, or races are intensifying at the moment. In fact, the existing economic inequality among population groups has not been resolved and is becoming rather worse. In other words, we are living in an age of uncertainty, where risk and opportunity coexist, because of the growing inequality.

Such uncertainty may be a harbinger of the coming revolutionary changes that will have a profound impact on the entire human race. Therefore, unless humankind exerts efforts to control or adjust to the rapid changes, along with efforts for the mitigation of the existing contradictions, the current crisis can bring humanity to an entirely uncontrollable contradictory society. Humanity, in fact, is presently standing at the crossroads of its destiny. If the control of change and the resolution of inequality are pursued in a desirable direction, human beings will be able to create an ideal community supported by the development of science and technology. If the conflicts and inequalities further deepen, however, without controlling the rapid changes, science and technology cannot be used in a desirable direction but will be used only to deepen the conflicts of interest among countries, religions, and races.

Industrialization and urbanization, which are the main characteristics of the modern society, are also bringing about rapid changes in ecosystems. Climate change and the destruction of ecosystems can cause civilization-level havoc when the global environment reaches a tipping point, where its resilience is lost.[92] Civilization began 10,000 years ago along with the shift of the cold climate to a warm climate after the atmospheric temperatures near the earth's surface increased by 5–6 degrees as the Ice Age retreated. Therefore, if current climate change occurs rapidly to an uncontrollable level, along with a rapid change in the ecosystem, humankind will be forced to face a crisis of having to survive in an entirely different global environment. The crisis at that time can be a civilization-level change whose size and content are difficult to predict now. The civilization-level havoc signifies that the human adaptation process needs to be required not only simply to the environment change but also to the political, economic, social, and cultural changes, which will be quite different from the current ones.

Humankind is already destroying ecosystems by abusing the earth's limited resources and raising the earth's surface temperature, causing uncertainty about the sustainability of the global environment itself. These global-scale changes may bring about catastrophic results unless we address them properly. Therefore, even if we will not be able to stop the change itself, we should at least be able to control the pace of change for our adjustment. As argued previously, the most important reason for the occurrence of chronic illnesses in the contemporary humans is that the adaptation process of the genes cannot keep up with the rate of lifestyle and living environmental change. If the rate of change is continuously fast or even accelerating, however, it can cause much serious problems leading to the crisis of human sustainability, way beyond health problems like the failure of chronic disease management and the epidemic of new diseases.

Can humankind greet a utopia where disease has been terminated?

Productivity refers to the amount of products or services produced when a unit of work is used, usually measured as the amount of output per work hour. The higher the productivity, the more the products are produced and the more easily the product becomes available. On the other hand, in a society with low productivity, products are so valuable that it is difficult for many people to obtain and use them easily. Until 10,000 years ago, our ancestors hunted and gathered food by using tools made of stone and wood, and because they had to rely on the plants or animals that are available in nature for their products, they formed small groups at best, living with such a low productivity. As the productivity increased with farming and herding since the beginning of civilization, humanity has been able to settle in certain areas, and expand their community as well. Further, with the Industrial Revolution, humankind made a dramatic advancement in boosting productivity. With the development of science and technology, the means of production were mechanized and automated to a considerable extent, while the nitrogen-based fertilizer increased the agricultural productivity, making it possible for common people to use products and obtain foods relatively easily and abundantly.

Therefore, human history can be interpreted as the history of ever-increasing productivity. If we solve our current conflicts and contradictions on the basis of productivity enhancement, we will be able to carve out an ideal future that is much better than the present. As the means of production become more automated in the future, the ability to produce more than the amount can be consumed by the entire human race will be realized soon. In addition, as the production volume becomes sufficient, the problem of distribution can be also solved, ushering in an age where everyone can use products freely anywhere. Of course, it is unlikely that a society running on the utopian principle of "production according to the ability and distribution according to the needs" as envisioned by many socialists, including Karl Marx, will come spontaneously any time soon. Despite the conflicts among many nations, ethnic groups, religions, and ideologies, however, the issue of production and distribution, which was the root cause of the conflict, can be controlled and managed appropriately before long.

If the time comes when the productivity is greatly improved and an appropriate level of production and distribution becomes technically feasible thanks to the development of science and technology, the community will be able to build a well-organized economic system that engages in production and distribution as necessary while ditching the legacy production relationship, where a few elites govern the majority. These production relations may again exert an influence on politics and the society, leading to the birth of a genuinely equal society, which will be formed via a union of free and independent individuals, which humankind has long dreamed of. In this imagined society, the safety and health of the members of the society will emerge as the most important social values, and the social security and medical system designed to protect them will become a fundamental social structure.

Aside from the aforementioned distribution of products according to individual necessity, in the utopian era, education will be provided according to people's talents and needs, free medical care will be provided for the treatment of all diseases, and conclusively, people will be able to do what they want. Moreover, as regional barriers among communities almost disappear, the boundary between states will likewise, disappear, and the political system can be transformed into a form of direct democracy, in which free individuals participate directly in the political decision-making process. The foundation of production may no longer be commoditized or enslaved labor but will be the labor aimed at realizing oneself, which signifies a transition to a community in which the monopolization of ownership will finally be dissolved. Labor can no longer be an act in the consciousness of labor for living, but an act of free will. Just as labor was not enslaved or commoditized but was a critical part of daily life in the age of hunter-gathers, labor will no longer exist for others but for oneself in the utopian era.

The productivity, mode of production, and structure of distribution are fundamental determinants of the political system, social structure, and cultural characteristics of the society. It can be said that the development of civilization in the past was largely determined or constrained by these factors. Therefore, if the production capacity exceeds the consumption and the dilemma of distribution is resolved, humankind will witness an era that is much advanced than the past politically, socially, and culturally. In particular, as the progress of science and medical technology leads to the eradication of disease, humans can survive well past the limits of their biological lifetimes, or they are able to control their lifespan in the future. In the end, humans will be able to welcome the utopian era they have been dreaming of for a long time.

Biological inequalities can lead to dystopia

Can humanity, however, greet such a utopian era in earnest? It may not be easy, even under the best scenario, for humankind to overcome the conflict and crisis that people are currently experiencing and to succeed in controlling and adjusting to change nicely, because even if we succeed in solving the many problems that humanity has experienced so far, new problems will certainly emerge in the society. The problems that will arise following the resolution of our current problems will be entirely new problems that we have never experienced before; further, they are the problems that may not be easy to solve, such as the aging of the population, the choice between life and death, and a stagnant society that has lost the vibrant energy for further development. In fact, unless special efforts are exerted, there is little chance that the future will evolve as we have imagined in the best scenario.

Now, the indicators of human conflict and crisis, such as income disparity, the unequal development of the society, and the differences in medical accessibility, are not decreasing but are even increasing. Therefore, the more likely scenario is that the productivity will improve without resolving contradiction in the production–distribution structure while science and medical technology continue to develop, thereby worsening the overall social and biological inequality. If only a limited number of groups will enjoy the achievements attainable through the advancement of science and medical technology, then only a subset of the people will acquire superior capabilities through enhanced biological functions, resulting in biological inequalities depending on social class. If such a scenario is ever realized, the future of humankind is likely to be a dystopia, not a utopia. Especially if such biological inequalities are realized, the future society will have crossed the point of no return because the dominance–subordination structure can be perpetuated, pushing humankind into irreconcilable conflicts.

Inequality is a major factor in the political, religious, and class conflicts faced by the humans living in the contemporary era. Looking back on the history of humankind, we can say that the people in the age of hunter-gatherers, which did not have any noticeable accumulation of wealth or class differentiation, belonged to an equal society. As civilization came into existence, however, communities like villages, towns, and cities were formed and then developed into nation-states before undergoing a series of changes to become unequal societies. As the power structure was created to manage and maintain the community, class division and wealth accumulation were realized, resulting in inequality among the members of the community. The class structure had been transformed from master–slave to citizen–noncitizen, nobility–commoner, and capitalist–worker according to the times. Fundamentally, however, we can say that the members of the society was divided into those who manage labor and own the surplus and those who barely own products to an extent that allow them to work and regenerate themselves.

Inequality occurs within communities like cities and countries, but it fundamentally arises from the differences in technology or productivity between communities. Even though various communities on the planet already showed considerable differences in their respective technological levels, the communities with superior technologies are becoming increasingly different in terms of application of science and technology from the other communities with comparably less advanced technologies. In a word, the inequality between communities is worsening in the future. In addition, if we keep the current trends, communities with advanced technologies can soon evolve to a point where they can apply their superior technology directly to the human body. Particularly if this inequality continues and extends to the point of the genetic manipulation of the embryonic cells, and the superior ability of only a few groups is transmitted to the their future generations, the age of evolution by natural selection may come to an end, and a new kind of human species that is superior to the current *H. sapiens* may be born.

No matter how large are the inequalities that are currently observed among countries, races, and groups, the differences are still limited to the opportunities for consumption and the cultural life, education, and healthcare access. The differences in physical and mental capacity, however, that may appear in the future societies may not be limited to the issue of mere opportunities. This difference will manifest as the difference between the new human beings that have been biologically strengthened through the enhancement of human abilities and the *H. sapiens* without such strengthening. Perhaps this difference may be greater than the difference between our ancestors who lived in the age of hunter-gatherers and the modern human who achieved civilization through the agricultural and industrial revolutions. It could be an absolute difference in capability between species, such as chimpanzees versus humans or, to a lesser extent, Neanderthals versus the *H. sapiens*.

Enhanced human abilities, another potential tool for domination

On the bright side, we will soon be able to fully understand the complex phenomena that cause and progress diseases on the basis of the power of the rapidly developing science. Based on such understanding, we will have a new future where the infectious and chronic diseases are rarely seen and the lifespan is greatly increased. Probably, the future over the next decades will be recorded as the time when the changes in the disease pattern happened most dramatically, and the increase of the human biological lifespan was the greatest in all ages. The change that will take place in the future, however, may be too fast and extensive to allow us to cope with, using the philosophical, ethical, and social concepts that humanity has established in the past, slowly over thousands of years. Therefore, it will be challenging for us to prepare for the future changes. Nevertheless, we, the present generation, cannot avoid our responsibility for the future, because the future society depends on us now, and it can give future humankind a more serious challenge if we do not prepare well now.

Since the age of hunter-gatherers, technological advances have transformed the tools that humans use, and newly emerged tools have led to the establishment and development of civilization. We are now entering an age where technological changes are transforming humankind itself. Just as it was difficult to imagine how civilization would evolve in the age of hunter-gatherers, however, it will be as difficult to anticipate the future with the view of present civilization. Just guessing, we may be able to find the key to unlocking age-old challenges like chronic illness and aging in the future, but it may also be the key to opening the door stepping into entirely new challenges at the same time.

Above all, as the share of the elderly population will be increasing in the future society, we will try to overcome the problems of aging by slowing down the aging process and maintaining our youth, or by strengthening human capability. For example, an elderly person with weakened muscles or who has difficulty walking can wear a human capability–strengthening device, such as a musculoskeletal assistant device, which will enable him/her to perform activities comparable to or more than those being performed by young people in their daily lives. Devices that enhance the biological ability to recognize and respond to things through memory or intelligence enhancement, and genetic engineering or gene expression control, as well as physical activity–boosting devices, will be developed and used in daily life widely.

Technology for strengthening the human body functions will be developed competitively and will first be applied to patients and the elderly population, but it can later be used to enhance human ability far beyond the average physical and mental capacity of the contemporary humans. For example, in some communities or groups, ordinary people, such as students, soldiers, and workers, other than patients or elderly people, may seek to gain superior status to people in the other communities or groups by using human capability enhancement technology. If we allow elderly people with weak muscles to live a healthy life with the use of a human capacity–boosting device, we will recognize such device as good one for helping weak and sick people. If a young and healthy person, however, enhances his/her human ability by using the device and defeats others in competitions, the device can be recognized as a tool for competition and domination.

As the group armed with the advanced weapons in the past conquered, enslaved, or extirpated groups armed with less advanced weapons, so too will the future group. Those who will have acquired stronger, faster, and healthier bodily capability as well as excellent mental capability in terms of memory or intelligence with the use of artificial intelligence devices try to dominate those without access to such strength-boosting devices. In the areas of learning, occupation, sports, and entertainment, people with enhanced physical and mental functions will dominate, and, in a severe case, a new relationship of dominance and subordination — the new master–slave relationship — can appear.

Unlike the natural selection process, however, in which superior genes are selected over a long period of time, the human body enhancement is a process of creating excellent physical and mental abilities artificially in a short period of time, suggesting that there is insufficient time for a variety of excellences to be verified extensively in terms of safety for humanity. Accordingly, such human capability enhancement process cannot be said to have been verified for securing our presence, nor it can be known what influence it would have on the sustainability of humankind. It is at this moment that the fate of humanity as *H. sapiens* becomes uncertain.

Stopping the tragedy of *H. sapiens* becoming slaves

Humanity ended up as the dominant species on earth not because it was able to use tools but because the cultural information produced by humanity was passed down and then accumulated. For instance, chimpanzees can use tools and can convey such use to the next generation in the form of behavior mimicry, but it is only humankind who can accumulate this information and transmit it in the form of thought. The accumulation and transmission of ideas by tools, such as language and script, were not only very efficient but also enabled the construction of a culture on which new ideas could be continuously added. It cannot be said, however, that the ability to create a culture has been equipped from the beginning of our hominid ancestors. As our ancestors began to eat meat and fish through hunting instead of simply eating plants and vegetables, humankind became omnivorous, thereby laying the foundation on which the brain could grow. The cerebral cortex has been developed more through the fierce pressure of natural selection for survival. Only after having their brains progressively advanced over a long period of time did the human race gradually develop the ability to produce culture.

In the past, the accumulation of information and the development of culture were based mainly on such ability of the neurons of the human brain, but now, far more information is stored in computers or information storage media than the amount that humans can accumulate in their brain. Now, humans and AI cannot be compared at all in terms of their storage capacity and information processing capability, suggesting that AI, not humans, will play a more important role in the creation and transmission of culture in the future. Until now, humans have played the role of the creator of culture because of their outstanding ability to see, touch, feel, think, and judge. The time will soon come, however, when AI takes over these senses and thinking and judging capabilities, or when people's abilities are enhanced far beyond the current human abilities with the help of AI.

As the framework of the community changes constantly and the roles of the computer keep growing, humanity will gradually depart from the cultural framework built on regional, racial, and religious foundations. Computer networks will provide a basis for new life so that people can meet and do virtually according to their common interests and tastes, without spatial constraints, and so that people will hardly feel the barriers of physical reality. The physical and mental functions that deteriorate with aging will also be supplemented or even strengthened by human enhancement tools and AI, which will again reduce the differences in cultural life according to the varying age. Furthermore, the traditional role of sex will be diminished as humans will have a stronger ability to control their sexual desire or cool alternatives to satisfy it. These changes will eventually transform today's family-centered society into a society based on new human relationships.

The signs of this unprecedented change are already envisaged today, and the wind of change will blow even stronger. If we have any choice at all amid such changes, the only available option for securing our future is that of controlling the speed and target subjects of the change. Thus, humanity's fate will depend on how fast we are going to change and how much control we can have on our target subjects. If the pace of change is very fast, the beneficiaries will soon turn out to be only the rich and powerful, and if the poor and the subordinates cannot have the opportunity for the change and will stay behind, the current inequality structure will further deepen. Depressingly, such a dark scenario is likely to be realized in the future. The rich and powerful people and groups will not only consolidate their current dominance but will also be able to strengthen themselves further and become excellent beings thanks to the advancements in science and medicine.

Perhaps the strengthening of the human capacities based on the desire to live forever can end up wiping out the civilization of the *H. sapiens* because new human species or superhominids can be born through the enhancement of human capacity. If we want to avoid the fate where the new human species become the ruling class and the *H. sapiens* become slaves, we must turn the direction of change toward ensuring the sustainability of humankind. This is also closely linked with ensuring harmony and balance among humanity, the ecosystem, and the global environment as well. It is because humankind's sustainability is only possible when the whole of humankind can keep pace with other life forms and the physicochemical constituents of our planet for the harmoniously changing world.

Chapter 5

After the End of Chronic Disease

5.1 Extension of Aging or of Youth?

During the past 150 years, the average life expectancy of humankind has jumped nearly two- or threefold. It is only the human race among all the species that has achieved a several-fold lifespan increase within such a short period. Such increase in the lifespan of humankind is basically due to the reduced mortality from disease. The improvement in one's lifestyle and living environment is likely to push the average life expectancy up to 100 years in the not-too-distant future. Ultimately we will enjoy long lifespan but face a new crisis at the same time as we will lose the driving force to inherit and develop traditional elements like family, society, and culture that have led human history through sex, descendants, and competition.

Eternal life, the elusive dream of humanity

A brief survey of the documents left by the early civilizations will reveal that humanity has long been interested in how long they can live as well as disease and death. An epic story from the Mesopotamian region features Gilgamesh, who was the king of Urk, one of the first cities. According to the epic, Gilgamesh was born a demigod, with two-thirds of his body being a god and the remaining one-third, human. As nobody could stand up to the powerful king, Gilgamesh took any woman he fancied to satisfy his sexual desire, regardless of whether she was the daughter of a soldier or the wife of his vassal. Upon hearing the angry protest of the people, the god created Enkidu, a barbarian who knew nothing about civilization and who would fight with King Gilgamesh. Enkidu went to Urk to fight with the king like an ox, but they soon became friends.

Enkidu died soon due to the wrath of the god, which brought deep sorrow to King Gilgamesh as well as a question about how long human can live. The king then went on a journey to meet Utnaphshtim, in search of eternal life. The only survivor of the Great Flood who was saved by building a ship, as ordered by the god, Utnaphshtim was given the gift of eternal life by the god. After going through some ups and downs, King Gilgamesh was able to obtain Ur-Shanabi from Utnaphshtim, an herb for rejuvenation. Sadly, he soon lost the herb to a snake, however, and had to accept his death in the end, after returning to Urk from his long journey.[93] The epic poem of Gilgamesh gives us the lesson that no one among us, not even the most superior humans, can avoid death.

The Changing Era of Diseases. https://doi.org/10.1016/B978-0-12-816439-6.00005-3

145

The search for eternal life was also one of the most pressing issues in ancient China. Emperor Qin Shi Huang assumed the throne at the age of 13 and then went on to build a powerful empire by conquering and uniting six countries at the age of 39. Even the mighty emperor, however, was afraid of death. He sent his envoys everywhere looking for an elixir of eternal life but with no result. He even sent ships carrying some 3000 boys and girls to the east for the same purpose. His envoys went to Jeju Island in Korea, and to Japan, in search of the elixir but failed to find it. Fearing execution for their failure to find the elixir, his envoys never came back to China. After finally accepting the cold reality that he would not be able to avoid death, the emperor ordered the building of a giant tomb (mausoleum) that could accommodate up to 10,000 people, in preparation for his life after death. The emperor eventually died at age 49 while on a journey to inspect the other regions of his empire. Ironically, his premature death was caused by mercury-laced medicine, which he ordered to be made to preserve his youth.[94]

Countless attempts have been made both in the east and west to avoid the fate of death, but nobody has succeeded in overcoming it, making the dream of eternal life humanity's ultimate pursuit yet to be realized. Not every living creature greets death as a matter of fact, however,; hydra and planarian flatworms seem to enjoy eternal life, without eventual death. Humanity, however, will never be able to avoid death as all mammals, including humans, are bound to die. The average life expectancy of the human species, however, has drastically changed from the time of hunters-gatherers, where the average life expectancy was 20–25, until the time of the contemporary society, where the life expectancy averages at 70–80 years. Therefore, humanity's long dream of extending human lifespan and maintaining one's youth may have been realized, at least partly, even though death is inevitable.

How long can we live?

Experimental studies have proved that fruit flies and rats live longer if they eat less. A phenomenon observed extensively from molds to apes, it can also be confirmed among humans to some extent. What then is the reason behind this phenomenon, which is commonly found among different species? That is, why do species live longer if they eat less? This may be related to the basic mechanism of the life phenomenon associated with survival and reproduction. If any species experiences famine due to the lack of available food, it is more advantageous for such species to use energy to sustain its survival rather than for its reproduction. The reason for this is that it is more effective for the species to restart reproductive activity for its breeding once it has survived the famine and when the food has been abundantly replenished.

Thus, cells might as well focus on repairing the damaged parts of their cells, allowing them to keep functioning normally and sustain their lives when their food is in short supply. As these activities are critical in preventing cells from falling into an abnormal state or changing into cancerous cells, they play a major role in preventing the cells from aging, or in preventing the occurrence of diseases, such as diabetes mellitus or cancer. To conclude, if you eat less, it will bring about the same effect as if you are experiencing a famine, prompting the mechanism of your body to realign itself in the direction of preserving your life rather than growing and reproducing itself. This not only reduces the incidences of disease but also slows down the progress of aging.

In 1988, David Friedman and Tom Johnson of the United States discovered the *age-1* gene mutation, which extends the lifespan of the roundworm (*Caenorhabditis elegans*), a nematode.[95] Shortly thereafter, gene mutations like *daf-2* and *daf-16* were additionally found, and subsequent animal studies showed that the lifespan increases if insulin-like growth factor-1, which is associated with such genes, scales back its overall activity.[96] Eventually, the genetic mutations that are known to be linked with a prolonged lifespan were found to be associated with the diminished role of insulin-like hormones. In the end, the increased experimental animals' lifespan was attributable to these genetic mutations, which played the role of making the cells' food supply insufficient.

In fact, if you eat too much, you are obliged to deal with the excess energy that will remain in your body after you consume the calorie that your body needs, forcing your body to receive a significant amount of oxidative stress from dealing with such an excess energy. Oxidative stress again impairs various normal functions of the cells, makes you likely suffer from chronic diseases, and shortens your lifespan. If the nutrients (e.g., sugar) in the cells are overflowing due to the oversupply of food, excessive efforts should be made to convert these into energy, suggesting that the mitochondria, which serve as energy power stations, are in a state of overworking, without enjoying any rest. Moreover, if the energy produced as such is not used appropriately, it will then be converted into fat and will be stored in the body. If such an imbalance between energy production and consumption persists, the mitochondria may reach their limit and stop functioning normally.

Let's say a product is made in the factory. If you make too many products at the factory or if the purchasing power of the consumer drops significantly, you will not be able to sell all the products you have made, thereby ending up storing the products in your warehouse. Even the warehouse, however, gets filled up before long and the factory will no longer be able to produce and thus will be forced to stop production. Likewise, when the energy produced by the mitochondria is not being consumed continuously, the mitochondria will lose its ability to work any longer. If the mitochondria, which are responsible for the production of energy for the use of human body, do not play their given role, the major tissues and cells of the human body will not work as they are supposed to, thus leading to diseases or shortening the lifespan. The mitochondria, however, become active again in energy production if the food intake decreases. In that case, the mitochondria will be given an opportunity to repair their malfunctioning parts either via autophagy or through their activity of fusion and division.

As such, the mitochondria will be regenerated, thus preventing disease or extending the person's lifespan.

Certainly, this effect cannot occur under the condition of excessive nutritional deficiency because the nutritional deficiency during a famine weakens the immune system and increases the likelihood of an infectious disease. In fact, when the periods of famine and those with no famine are compared, it can be seen that the mortality rate of children born just before or during a famine was generally higher because the nutritional deficiency weakened the immune system.[97] Therefore, the lifespan can be increased only if the food intake is limited but not to a level too low as to cause malnutrition. In the end, we need to eat a variety of nutrients, such as ones from vegetables, fruits, and nuts, while maintaining a balanced diet of carbohydrates, proteins, and fats, but at the same time reducing our overall food intake to achieve the desired results.

After analyzing six large-scale studies that had tracked their subjects for more than a decade, Stephen Moore of Harvard University found that a 75-min brisk walk per week increased the subjects' lifespan by 1.8 years and that a 30-min workout for 5 days a week increased the lifespan by up to 4.5 years.[98] A recalculation of these results will show that exercise will have the effect of extending the lifespan by more than nine times the total exercise time we consume. It is such a profitable investment! Exercise not only improves the cardiovascular function but also improves the quality of life. In addition, it alleviates tension and reduces negative emotions like anxiety, depression, and anger while enhancing the mental activity and improving the memory, thereby helping prevent mental disorders like Alzheimer disease among the elderly. On the other hand, 1%–2% of the total oxygen used by the cell is converted to reactive oxygen species, which is essential for signal transmission within cells. Excessive exercise, however, increases the oxygen consumption and produces more reactive oxygen species than the appropriate amount, allowing the reactive oxygen species to attack the DNA, RNA, and protein structures in the cell, thereby slowing down the cellular function and accelerating the aging process. Therefore, it is crucial to exercise regularly but not excessively.

Humanity is allegedly known to have begun fermenting and drinking wines with the dawn of civilization, but humans had already been consuming alcohol contained in foods for a long time before that. Therefore, alcohol is not a new chemical substance unfamiliar to our bodies. As the alcohol consumed before the advent of civilization, however, was made from the natural fermentation of food, it was not enough to allow our ancestors to ingest a lot of alcohol, as we do today. On the other hand, those who were able to eat and use aged and fermented foods could have benefited from the wide range of foods that they could consume compared with those who could not eat fermented food at all. The pressure of this natural selection helped humankind to adapt to a small amount of alcohol intake today.

Therefore, alcohol has a beneficial effect on health and increases the lifespan to a certain extent. People who drink a glass of beer every day can reduce their mortality from cardiovascular disorder and can increase their lifespan by more than 2 years. Moreover, if we drink a glass of red wine a day, our lifespan will increase by up to 5 years.[99] An examination of 34 large-scale prospective observational studies on the relationship between alcohol consumption and mortality revealed that alcohol consumption by up to 6 g a day, or about half a glass, could lower mortality by up to 19%. Of course, moderate drinking is very important because consuming more than two glasses of alcoholic beverage a day can again raise the mortality rate.[100]

An article in the *New England Journal of Medicine* in 2013 sought to determine the extent to which one's lifespan would increase if one quits smoking. The study researcher, Dr. Prabhat Jha, found that people who smoke would die 10 years sooner than those who do not, but if they quit smoking at a young age (between 25 and 34), they could live as long as those who have never smoked. Although the lifespan lengthening effect diminishes as the time of quitting smoking is pushed back, it was discovered that even if one stops smoking at a somewhat late age (55–64), he or she is likely to live 4 years longer than those who do not quit smoking at that age.[101] Meanwhile, Dr. Michael Thun's study published in the same journal showed that the mortality rate of women with lung cancer or chronic obstructive pulmonary disease, which is closely related to tobacco smoking, is approaching the mortality rate of their male counterparts. This is a phenomenon that has occurred since the 1960s, when the smoking rate of men began to fall, whereas that of women rose. In other words, if people smoke, the mortality rate increases and the lifespan is shortened regardless of gender.[102]

Taken together, these results suggest that a healthy lifestyle, such as regular exercise, moderate drinking, and smoking cessation, can increase the lifespan by more than 20 years. If we reduce the amount of chemicals we are exposed to via air pollution and our living environment, as well as the level of various stresses we go through, our life expectancy will increase even more. Given that the average life expectancy in developed countries is currently around 80 years, the improvement of one's lifestyle and living environment is likely to push the average life expectancy further up to 100 years in the not-too-distant future.

Markedly increased warranty period for life

In the end, we will be able to enter the era of 100-year life expectancy by improving our lifestyle and living environment. Is the 100-year lifespan, however, the biological limit of life imposed on humans? As it is unprecedented that the life expectancy of humankind has increased so much, the scientific basis of how long we can live is not clear at all. It is reasonable to assume, however, that the biological limit of human life is at least 100 years given that there are people who live beyond the age of 100.

Why then is there a limit in the human lifespan, and how is the limit determined? The most important purpose of survival for all life forms, including humans, is to give birth to offspring and to transmit their genetic information to their posterity. If humans are not destined to die, it will not be necessary for them to transmit their genetic information to their posterity, rendering the birth of offspring an obsolete goal of survival. Because we are all destined to die, the transmission of genes to our offspring is the most important goal of our survival. In order to achieve the purpose of survival, it is necessary to survive until not only we give birth to our offspring but also until our offspring give birth to the next generation, suggesting that there is a warranty period for life. That is, humans should not die until it is ensured that their genetic information will be passed down well to their future generations.

The level of assurance, however, may vary depending on the lifecycle stage. At least until one succeeds in giving birth to the second generation, there must be a definite guarantee, and until the birth of the third generation by mating of the second generation, there must be a certain degree of guarantee, if not as sure a guarantee as was given for the birth of the second generation. Then, it can be ensured that the genetic information has been passed down well to the posterity. When the third generation is born, however, the responsibility and authority are transferred to the second generation, and the purpose of survival of the first generation is no longer clear, with virtually no guarantee of further life. In fact, aging can be said to be a phenomenon in which the formerly active life begins to be weakened since the expiration of the definite warranty period, which was essentially necessary to produce and raise offspring.

The length of the warranty for life is also affected by the probability of death (a chance of dying). The higher the mortality rate, the earlier the offspring will be born due to the urgent pressure to produce offspring sooner, with a corresponding reduction in the warranty period for the entire life. In rats, the mortality rate is high because they are easily eaten by other animals, which again shortens the time until their sexual maturity. Therefore, the warranty period of life is also short. If the probability of death by being eaten is mitigated, however, along with the falling mortality rates, the warranty period of life also changes. Ever since the North American rats began to live on previously uninhabited islands where predators are much less, their lifespan increased by more than 50%; furthermore, their aging rates have been reduced at a rate inversely proportional to their increasing lifespan.[103] Moreover, as the Galapagos tortoise can survive any attack thanks to its thick back cover, its mortality rate is not high, and the period until its sexual maturity is long enough to enable it to produce sperms only after they are 35 years old. Thanks to this, the time to give birth to its offspring is extended as well, and the warranty period of its life has also been extended. The Galapagos tortoise is generally known to live for well over 100 years.

The warranty period of life is related to the metabolic rates as well. In mammals, the larger the body size, the lower the metabolic rate and the longer the lifespan, while the smaller the body size, the higher the metabolic rate and the shorter the lifespan. For example, the metabolic rate of rats is seven times higher than that of humans, whereas their lifespan is much shorter than that of humans.

A high metabolic rate is equivalent to a high RPM of automobile engines, suggesting that if the engine is used a lot, the lifetime of the engine becomes shorter. Likewise, if the metabolic rate is high, the warranty period of life will also be shortened. Interestingly, however, the lifespan of birds and mammals varies greatly even though they share the same metabolic rates. A comparison of the lifespans of pigeons and mice, which have similar metabolic rates, will reveal that the lifespan of pigeons is about 35 years while that of rats is only 3–4 years.[104] The difference is nearly about 10 times. This difference may be due to the fact that oxidative stress is much less likely to occur in birds than in mammals. More importantly, however, the warranty period of life is greatly influenced by the probability of death. In other words, pigeons are less likely to be eaten by predators than are mice.

Telomeres are also known to be related with the lifespan. In the 1930s, Barbara McClintock and Hermann Muller argued that the repetitive nucleotide sequences at the ends of chromosomes play an important role in maintaining the stability of the chromosomes by preventing the chromosomes from being worn out or sticking to other chromosomes. Muller called such structures "telomeres," a name coined by combining *telos*, which means the "end of a structure," to signify the lidlike attachment at the end of the chromosome, and *meros*, which means "a part."[105] Telomeres are thought to be related with the aging or lifespan because they tend to become shorter as cells become differentiated or older. It is not yet clear, however, whether telomeres are a direct cause of aging or a determinant of a person's lifespan, or are only among the telling signs of aging.

Through the aforementioned studies on molecular biology, we gained a deeper knowledge on biological phenomena, but it is too early to say that we have clearly understood the factors that determine the phenomenon of aging or lifespan. Understanding the factors affecting a person's lifespan requires an observation of the macroscopic changes as well as of those occurring at the cellular or genetic level. The mortality rate of humans has fallen dramatically, and the time to produce offspring comes much later than in the past, while the overall time required to raise one's offspring is also being extended. During the past 150 years, the average life expectancy of humankind has nearly doubled or even tripled. The several-fold extension of the human lifespan observed in this short period of time is fairly unique among all biological species. Why is this happening? The factors known to affect human longevity, such as the genes, metabolic rates, and telomeres, have not changed dramatically over the last 150 years In fact, it is not possible for them to change in such a short period! Ultimately, it can be said that the remarkable increase in the human lifespan is basically due to the reduction of the mortality rate from disease.

Extension of aging or youth

Many studies on longevity and aging have revealed that when the lifespan increases, the impediments to physical activity or disabilities are delayed as much.

In other words, not only the lifespan is stretched, but the healthy period during the lifespan is extended as well. Perhaps there is a biological clock imprinted on the genes in our body that controls our lifespan and aging. This biological clock divides the lifespan into the maturation phase and the aging phase. The longer the lifespan, the longer both the maturation and aging phases would become. The aging phase can be understood as the period of living with the functions and abilities one obtained during the maturation phase.[106]

Therefore, if the lifespan is increased up to 100 years, the age at which aging begins will be pushed back significantly compared with the present beginning of aging. In modern societies with the average life expectancy reaching up to 80 years, it can be seen from the observation of the current elderly generations that the aging process has been delayed compared with some 50 years ago, when the average life expectancy was about 50 years. Likewise, when the average life expectancy reaches 100 years, the aging process will begin much later than it does today. Therefore, humans will enjoy a most active life at 50, before entering into the aging process. By that time, the standard age of elderly people, currently pegged at 60 or 65, will no longer be considered the standard age for elderly people. As the aging process is being delayed, persons aged over 80 or 85 may be considered the elderly people who can no longer actively engage in social activities.

Indeed, studies on the elderly population have shown that impediments to physical activity are less common among the elderly people today than they were among those in the same age group in the past. The disabilities that frequently appeared before people turned 65 in the past are now occurring in the population older than 65 years. This phenomenon can be attributed not only to the extended healthy period of lifespan and the advancement of medical management, such as early diagnosis and effective treatment, but also to the improvements in the general living environment, such as the improved residential facilities and the convenient public transportation, as well as to the improved education and living standards.

A longer life expectancy has long been a dream of humankind, but as a result of lengthened lifespan, the absolute number and proportion of the elderly population in the total population has greatly increased. Can this be considered a success when viewed from the perspective of realizing a healthy society? Those who see this as a failure say that the number of people with diseases or disabilities is increasing with more people reaching the elderly age. As it looms large as a huge social burden, an extended life expectancy may result in a painful failure rather than a rosy dream when you look at the society as a whole.

On the other hand, those who see it as a success argue that the share of people with a disease or disability does not increase even if the aged population increases because it is generally healthy persons who reach an elderly age. In fact, in a study of very old people (older than 110 years) living in the United States, about 40% of them were reported to still be enjoying independent lives. Their functional level was also not significantly different from that of the elderly people who were between 92 and 100 years old, suggesting that medical expenses of the elderly will not necessarily go up even if humanity enters the super-aged society.[107] If this "success" hypothesis is true, healthy persons will grow older without diseases, increasing the elderly population without necessarily increasing medical expenses. Even though a decline in bodily function will still accompany aging, the incidences of disability and disease, which are potentially great social burdens, will not necessarily increase by as much.

On the other hand, a Greek myth on aging shows how lucky we are because we are bound to die considering that aging is inevitable as our lifespan increases. One day, Eos, the Goddess of the Dawn, who was born with eternal youth, fell in love with Tithonos, the Prince of Troy, who was very handsome. One day she went to Zeus and begged him to let Tithonos live eternally, without ever dying. In the end, Tithonos, who became immortal because Zeus granted Eos's wish, did not die but continued to grow older because she did not request to Zeus that Tithonos keeps youth forever. It is a sad story because Tithonos, looking at Eos who is young forever as a goddess, had had to live in bitterness.

Perhaps the day will soon come when elderly people will account for the majority of the human population as the life expectancy is extended up to the human biological limit and the mortality rate falls sharply with the rapid advancement of science and technology and medical care. Moreover, if the lifespan extends beyond the human biological limits through human enhancement and regeneration, humankind will enjoy an even longer lifespan. However, when such extended lifespan is realized, competition for natural selection from the desire of sex and descendants is diminishing. When this happens, humanity may face a new crisis as humankind will lose the driving force to inherit and develop traditional elements like family, society, and culture that have led human history. Therefore, while finding a way to extend the lifespan and to slow down the aging process is a task that should be carried out by humankind in the future, promoting an environment where death can be met without tragedy at a certain age will be another crucial task that merits attention not to lose the driving force of humanity.

5.2 Another Crisis of Humanity Looms Large With the End of Disease

Perhaps the world population will have peaked in the 21st century, because it can be predicted that the population will undergo a downturn due to the declining birth rate once it reaches the peak. Along with these changes, each individual in the family becomes more dependent on the social network system than on the family relationship. Therefore, the kinship among the family members will get thinner. On the other hand, the social network system will be able to control individual judgments and even emotions. Here, absolute power that one can trust and rely on more than anyone else can be born. Therefore, the development of medical technology that can bring about the end of disease has an inherent potential to cause other problems. It is time for humankind to begin efforts to prevent such changes from escalating into a crisis of humanity.

Decreased mortality lowers the overall fertility rates

When humans migrated out of Africa and dispersed to the four corners of the world to reside there, the hunter-gatherer population rose to about 50,000, after which it grew to 5 million before the beginning of the Agricultural Revolution.[87] As the population growth at this time was not due to the change of productivity but to the expansion of the overall residential area, it can be said that the population grew purely depending on hunting and gathering. In other words, probably 5 million is the maximum number of people who can live in the natural environment of planet Earth without special skills and tools. The extent to which the population can increase more than the number that can be supported by hunting and gathering, however, is ultimately decided by how much additional productivity can be gained. This is because productivity affects the socioeconomic relations and quality of life and thus affects the health status as well.

In the age of hunter-gatherers, there was no means of transportation other than walking or running, and there was a limit to the distance traveled because the booty acquired through hunting or gathering had to be moved by bare hands. In other words, the productivity was limited because there was no faculty or means to get enough food. As the limited productivity of the hunter-gatherers makes it difficult to maintain the herd population, the low productivity in the age of hunter-gatherers was the critical factor that kept the population growth relatively constant.[4] Moreover, the hunter-gatherers often had to switch their habitats to secure a hunting and scavenging area where they could maintain a decent life. The frequent movement of residences also made child birth and rearing difficult, making it another factor in suppressing the population growth.

Although population growth is largely determined by the difference between the birth and death rates, the underlying factors that influence fertility and mortality are the productivity of the region. If the productivity is low, the population cannot increase, but if the productivity increases, the population can increase as well to keep pace with the increased productivity. Probably, the first consequence of the increase in productivity would be the increase in the fertility rate. Also, with the increasing productivity, which again leads to a more affluent life and a higher standard of living, a change occurs in the overall mortality rate. The population was rapidly increasing during this period due to the increase in the fertility rate and the decrease in the mortality rate. On the other hand, a decrease in mortality means that the probability of death has been reduced. If the probability of death is lowered, the desire to produce offspring also falls, and the fertility rate decreases again after a certain period of time. As such, the continuing increase in productivity does not always result in a continuous increase in population.

After the Industrial Revolution, the birth rate increased along with the rising productivity. After the mid-19th century, the mortality rate began to fall, and the population began to increase dramatically as the difference between the birth and mortality rates enlarged. As industrialization progressed not only in developed countries but also in many other countries beginning in the 20th century, population growth became a global phenomenon. In low-income countries, in particular, the increase in the birth rate and the decrease in the death rate triggered an explosive population growth. By the beginning of 21st century, however, most developed countries had begun experiencing a declining fertility rate following a decline in their mortality rates. Perhaps the level of health care in the future societies that would boast higher productivity than today's modern societies will be much higher than now, and the mortality rate will be correspondingly lower. If the mortality rate is lowered such, the probability of death will drop as well, pushing down the fertility rate further before long. Therefore, even if the population increases globally in the foreseeable future, it can be predicted that the population will undergo a downturn due to the declining birth rate once it reaches the peak. Perhaps the world population will have peaked in the 21st century.

Increasing lifespan could abolish the traditional family relationships

Beginning from the age of hunter-gatherers to the arrival of the Industrial Revolution, the population gradually increased, but there was no significant increase in humanity's lifespan because the mortality rate did not decline significantly despite the rising birth rate. There were a few people who survived until they passed the age of 40 only by luck. From the middle of the 19th century, however, the mortality rate had begun to decline in earnest, triggering a marked increase in the average human lifespan. In Sweden, where fairly accurate records have been kept on the country's population, the life expectancy of a person born in 1840 was 45 years for women, but it has now risen to above 83 years.[108] Such change means that the average lifespan has increased by 3 months every year. People everywhere, not only in Sweden but also in Western Europe and North America, experienced almost the same increase in lifespan. The average lifespan has increased at even steeper rates recently in Asia, where the social development has been lagging behind that in Western Europe and North America. In countries like South Korea, where the lifespan has increased in earnest during the past 60 years, the average lifespan has increased by almost 6 months every year. The increase in longevity is attributable largely to the improvement in medical technology, such as antibiotics and vaccine development, but fundamentally, it is also attributable to the abundant food and hygiene environment brought about by the Industrial Revolution or equivalent industrialization.

Of course, an increase in lifespan has long been a dream of humankind, but as human life expectancy is expected to increase further in the future, it is time to consider the various changes that the growing lifespan can bring about.

In fact, the increase in lifespan can lead to a serious challenge rather than a rosy future. First, as the human lifespan increases, the fertility rate decreases as discussed here earlier. The reasons for the declining fertility rate may include increase in the social activities of women, increase in the period of education, delaying marriage, and the burden of child rearing. Fundamentally, however, the increasing life expectancy and the decreasing probability of death are playing the most crucial roles in suppressing the motivation for childbirth. Indeed, the average lifespan has increased by 20–30 years in most countries during the past 50 years, with the fertility declining by 30%–60%. The fact that the women in countries with life expectancy of 50–60 years gave birth to five to six children, whereas most of the people in countries with life expectancy of 80 years or over had fewer than three children shows an inverse relationship between life expectancy and birth rates.

As life expectancy increases further, the fertility rate may perhaps fall close to 1. In other words, people will increasingly tend to have one child. Basically, the change in the fertility rate is largely due to the change in the mortality rate. As the probability of passing down one's genes to the next generation through one child is reduced when the mortality rate is high, people tend to have many children to increase the probability that their genes will be passed down to their offspring. On the other hand, the fertility rate will be low if the mortality rate is very low because the genes can be securely passed down to the offspring even via one child. The reason that the number of babies born to couples tends to fall from more than one to only one is that the genes can be transmitted anyway to their offspring from the perspective of each couple, even in the case of one child. As the couples have achieved the basic life purpose of transmitting their genes, the reason for having two or more children has also been disappearing.

An increase in life expectancy also means that married couples have to live together for a very long period of time after their children have grown up. In addition, only one child means that the share of parenting period in the marriage life is reduced by as much. Furthermore, the very meaning of the marital relationship will change as well because the nonreproductive period becomes longer than the reproductive period in their marriage life. In other words, the relationship as one's lifetime partner will be valued more than the traditional marital relationship founded on sex. Therefore, various forms of family, including diverse forms of heterosexual cohabitations, homosexual couples, and group residential communities, can appear more frequently. For example, let's say that a couple has now turned 60 years old, and their children have already left them. If their life expectancy is 80 years, they are more likely to keep their marriage. If they live past 100 years, however, they are less likely to maintain their marriage until their death because they have to grapple with more than 40 additional years of nonreproductive life. If the healthy lifespan increases along with the increasing life expectancy, thereby enabling people to engage in social activities for a considerable period even in the old age, the likelihood of meeting new mates and resuming marriage or cohabitation with them will rise as well.

Family life, including the size of family, has undergone many ups and downs since the period before the Industrial Revolution, when humans strove to ensure the delivery of their genetic information down to their descendants by raising the birth rate because the death rate was high. In the contemporary era, however, the birth rate began to drop along with the falling mortality in most of developed countries. In the years before the Industrial Revolution or equivalent industrialization, people living with large families enjoyed intimate relationships with each other, but the family size is getting smaller in the modern society, and the familial kinship is thinning as well. If couples bear only one child and each individual in the family becomes more dependent on the social network system than on the family relationship, along with a markedly growing lifespan in the future, the kinship among the family members will get even thinner. In other words, each individual will identify himself or herself much more significantly as a member of the society than as a family member, as the conventional family life will lose ground over time.

Construction of a new community more attuned to the network society

In the future societies, work or education will not be carried out in a specific place, such as in the workplace or in school, but in one's residence, by logging into a computer network. Such change can transform one's habitat into an important place interfacing with an entire community. Therefore, the place of residence will no longer be merely a unit of family but will evolve into a new community that is different from a traditional family. The increase in the cases of celibacy, separation, divorce and remarriage, and same-sex marriage in recent years can be considered the beginning of new familial relationships. In addition, various types of experimental residential communities existing today may be seen as indicative of a search for creating a community that will emerge as the center of the future society.

As the mortality rate declines and the birth rate falls, the proportion of elderly people over 65 in the future society will rise sharply. If the majority of the population becomes old, however, all the elements of civilization, such as the society, culture, education, and politics, as well as the nature of medical care, can also be influenced and changed. That is, the history of humankind, which has been driven by sex and power, can go into a stagnant phase by gradually losing its driving force in the society dominated by the elderly population. Eventually, it will be human enhancement tools, artificial intelligence (AI), or robots that will keep such stagnant society running. Although human enhancement tools will reinforce the physical and mental abilities of humans and support human activities, the fundamental forces to drive the community are AI and robots that will directly produce commodity and provide services. AI and robots will replace much of the labor, and the days when humans are the subjects doing labor will soon be gone.

The human society has been increasing its size from the precivilization era until the present day—that is, from small groups of people in the early days to the present nation-state or global community. The basic unit, however, remains the same: the family. During the age of hunter-gatherers, the sense of belonging to the family was so great that the individuals' self-consciousness could not be distinguished from the family identity. However, such kinship among the family members gradually faded as people progressively developed a sense of belonging to those communities greater than the family, as the size of the community began to increase. In the future, the transition from a local or national community to a global community, and the emergence of diverse forms of family, will weaken the traditional family bond even more. This change also suggests that the medical care will shift from the traditional family-centered care to the care provided by the social network system composed of AI, robots, and medical professionals. As the family has been the basic community unit from the time of our prehistoric ancestors or the premodern society until now, if it loses its traditional role, each person will lose his/her absolutely trustworthy group.

On the other hand, although the family is a social unit, its formation may be attributed to the biological characteristics of humans, such as the sexual differentiation and the birth and rearing of offspring. Family based on sex and reproduction is observed not only in humans but also in other animals, suggesting that the family unit is the product of evolution devised for more efficient adaptation to the environment. Therefore, no matter how great the social change is, biological humans cannot completely abandon such essential characteristics that evolution has created. Perhaps even if the meaning of the "traditional family" changes, the family will remain in some other form in the *Homo sapiens* society as long as the baby is born to the biological mother, not born in the factory, as suggested by Aldous Huxley in his novel *Brave New World*.[109] In the end, it will be a challenge for the future society to create a new community that will be capable of smooth adaptation to the future changes while at the same time maintaining a feeling of belonging to the family or a sense of kinship among family members.

Emergence of invisible absolute power

In the future society, the power of the nation-state or government founded on race, ethnicity, and religion will gradually disappear, although the legislative, judicial, and administrative functions of communities may remain somehow. Individuals will no longer be regarded as members of a nation-state but will become members of the global community or the community of humankind. Leading these shifting trends is the increase in information owned by individuals, because the more information an individual has, the less the amount of information held exclusively by the authority, and the less the control exercised over individuals. The increase in the amount of information held by individuals is often accompanied by a demand for changes in the political and social relations.

The revolutionary changes in civil and political rights that began in Europe in the 17th century, such as the civil revolution, the development of modern democracy, the collapse of the monarchy and dictatorship, as well as the Renaissance of the 14th-century Europe, were all brought about by the increase in the volume of information held by individuals. This can be seen in the case of the large number of Bible prints made available in Europe by Gutenberg's metal letterpress printing machine, which threatened the vested power of the religious authorities and priests. With the growing spread of the Bible among the general public, a new religion (Protestantism) emerged, which then became the engine that propelled the emergence of a new society. Looking back at humankind's history, we can see that the society advanced step by step in the wake of the weakening control over individuals, but the instability inherent in such changes often led to wars and revolutions, in some cases, acting reactively, laying the groundwork for the emergence of a violent form of power, such as totalitarianism or dictatorship.

Production relations will also show an entirely new pattern in the future, along with a rising productivity different from that which humankind had been familiar with in the past. As the computerized network forms the basis of the production relations, the individuals or groups that effectively use such production relation will lead the society. The relationship between the dominant and the subordinates will not be clear in the society, just as no such relationship can be seen in the Internet. When the amount of information held by individuals increase, it can be said that a new invisible power that collects and organizes information will emerge because of the individual's increasing dependence on the information provision system. Invisible absolute power such as Big Brother in George Orwell's *1984* may not be revealed openly to the public, but there are certainly individuals or groups leading the network system.[110] Such individuals or groups will probably not be defined clearly as dominant classes as we observed in the past because they are constantly changing rather than specifically defined individuals or groups, such as reigning leaders or ruling classes. They can be understood, however, as a new ruling class that exercises power over people through network systems.

Unlike the traditional political powers in history, these powers will have no borders or regions and will not be physically violent, but they will be able to control individual judgments and even emotions and place each individual in a position of absolute subordination. For example, let's say you are a mid-30s man who had been employed in a company after graduating from college. All the information related to your marriage, family life, residence, commuting, meal pattern, co-workers, exercise, book reading, travel, etc. is gathered through a computerized network system, and is tabulated and analyzed, and then the information useful in your life is provided by the system. Based on your information, for instance, you are expected to be in a state where your blood sugar will go beyond the normal level after 5 years, and you will need medication after 10 years. Further measures to prevent this development will be provided according to your genetic and bio-monitoring information. As these results are critical information required to prevent disease, you may feel grateful and will become more dependent on the system that provides this information. Therefore, you will end up believing that this system is reliable and secure and that it understands you and reflects your needs. Here, absolute power that one can trust and rely on more than anyone else is born. The problem is that if a certain group gains exclusive control over this invisible power, or if the system's core is an AI that cannot be controlled, humankind will be inevitably forced to leave its fate to such invisible power.

The end of current diseases can invite new diseases

The era of infectious diseases began after the agricultural revolution took place, a time when the community began to increase in size and live close to animals by farming and herding. The age of chronic diseases following the Industrial Revolution can be said to have been caused by increased caloric intake and by the growing number of factors detrimental to human health, such as smoking, exposure to chemicals, and stress, in the wake of the drastic change in humanity's lifestyle. Accordingly, we can say that the pattern of disease is basically determined by the circumstances of the time. The changes that have already started in the contemporary age are increase of the human lifespan, along with a decrease in the fertility rate, an increase in the elderly population, and the weakening of binding power of the family. This shift will change not only the man-man relationship but also the man-machine relationship, thereby evolving into a relationship that is totally different from the past.

The modern society is in the process of undergoing such changes. The successful prevention and treatment of the current chronic diseases will further accelerate the aforementioned changes, and as we have seen in the past civilizations and in the history of diseases, the sooner the society changes, the more likely it is that physical, mental, and social maladjustment will occur, opening up a new era of disease. In the end, to address the contradictory situation in which a new epidemic can occur when the current one is over, it is not enough to simply accept the changes because if the current change will be further accelerated, the possibility of the emergence of a new problem or disease will increase as much.

Therefore, the development of medical technology that can bring about the end of disease has an inherent potential to cause other problems. What will happen if the human capacity enhancement device will allow us to have excellent abilities, and if the strengthening of human abilities will be realized in a discriminatory way based on wealth, power, or specific population groups? Probably, we will no longer accept human abilities as they were given by nature; on the contrary, we will reveal the desire to have superior abilities through the help of such devices, to win the competition with others, and to live longer and healthier lives compared to others. This will undermine the biological laws that have so far relied on natural selection, as well as the moral and ethical foundations of humanity.

People have accepted their given biological abilities and have been living with the conditions that they have received from their parents, but they will likely no longer accept their existence in current form if they can develop superior abilities using artificial means. Competition will also depend more on the abilities enhanced by the strengthening device of the human body than on the natural abilities, which means that we will see greater dependence on machines and devices. Eventually, humanity will gradually transfer its very nature to machines and devices. These changes may lead to problems like identity crisis, loss of self-worth, adjustment disorder, and depression, opening up a new age of illnesses: the Age of Mental Illnesses. Therefore, humanity is now at a critical juncture. It is time for humankind to begin efforts in earnest to prevent such changes from escalating into a crisis of humanity.

5.3 Pondering Life and Death

We are progressively approaching to death while going through the process of growth and aging. As the amount of muscle is reduced and the cardiac and brain functions fall, people eventually face death. However, the death of individual humans may be a necessary program to maintain the vitality of the entire human race. If humankind acquires everlasting life, any further advancement of human life will be halted because eternal life can no longer evolve further after it loses the motivation to compete for natural selection. Therefore, eternal life will fall into the paradox of loss of vitality. In fact, death is the door to a new life, which will assure the continuation of humanity well into the next generation.

Life changes toward death

It is now time to think about the most essential issues of human, the life and death, and to renew the meaning of human existence and to think about the correct application of future medical technology. We have to look back at history in the present age of uncertainty because when we are just trapped in the current phenomenon, it is difficult for us to have an accurate overview of what has happened with the flow of time. From the very beginning, the universe underwent a series of complex changes within a period of 13.5 billion years, and such changes created the basis for the existence of planets like Earth, and of living organisms, such as humans. Moreover, the changes never stop but are still happening today. We cannot clearly see where the changes are ultimately heading, and what their purpose is. The only thing we can be certain of is that everything, including living organisms, is changing.

It is estimated that the first life form on earth appeared some 1 billion years after the birth of Earth, or about 3.5 billion years ago. The first creature was probably a simple form of nucleic acid formed by the combination of small molecules. This first creature born from an inanimate object must have had an amazing ability to replicate itself. In other words, it must have been the origin of the living organisms as we know them today, which has transmitted their genetic information through reproduction. It seems that there is no way to find out exactly how life began, but the first creature must have given birth to descendants by passing down the genetic material of its nucleic acid to the next generation, and such single life form has been the origin of all the living organisms on earth and have evolved into humans after a long time.

Inanimate objects have also changed constantly along with age of the earth over time. As the earth's crust shifted, the continents began to move, forming islands and mountains in its wake. Inanimate matter, such as air or soil, was formed, heavily influenced by the activities of living organisms. What distinguishes living organisms from inanimate objects, however, is that the former produce descendants who are like themselves. This is so because a very important mission has been given to life forms: to convey to their descendants the key information made up of the nucleic acid, called "DNA." These nucleic acids have been competing with one another to create a variety of life forms. Those among them that had evolved into a favored form of survival by better adapting to the environment came to occupy a dominant position and thus succeeded in spreading a greater number of offspring. In other words, natural selection was the underlying force that brought about the changes.

Richard Dawkins argued that the purpose of life is to transmit genes composed of DNA, which can be deemed as the framework of life, to the descendants.[111] According to Dawkins, humans exist for the purpose of passing on their genes to their descendants. The process of transferring one's genes to one's offspring, however, is equal to the creation of offspring equipped with a new genetic combination by competitively selecting a partner among the population. Therefore, once a person has achieved the purpose of his/her existence through successful genetic transfer, he/she will no longer have any reason to live more and will thus inevitably succumb to death. If the lifespan of living organisms is not limited, however, there will be no reason for each living organism to pass on its genes to its descendants via a competitive process. In other words, the delivery of one's genes within one's limited lifetime is the prime purpose of human existence. If so, isn't death or a limited lifespan indispensable for achieving the foremost purpose of human life?

Another key aspect of life is that it constantly changes throughout its existence. What I am today cannot be what I will be tomorrow, and there is no such thing as the absoluteness of our existence, or some property of myself that never changes. I cannot even be assured of still existing tomorrow, or I will not be biologically the same tomorrow as I am today even though my continued existence tomorrow is guaranteed, because many of my cells will surely change tomorrow, and my cells' functions will not be the same. As human relations have changed socially along with the changes in the environment, the ecological position of human existence has also changed. In the end, we are constantly changing and are progressively approaching to death while going through the process of growth and aging. Therefore, life must be understood as the process of growth, aging, and death. A fixed life or existence is only a fictional concept. The process of such relentless change must also be understood in the context of the flow of time. Time is not something in which you can undo later what you did once. That is, the changes undergone by living organisms cannot be reversed again along the order of birth, growth, aging, and death. What we can only do, if we can, is to slow down the process of change that is bound to take place along the flow of time.

Death: a device to sustain life

A new history might be recorded in this century when diseases disappear and the biological strengthening of humanity further proceeds. The transformation into an excellent being that has overcome disease and death has long been a dream of humankind. This dream indicates humankind's aspiration to go beyond the historical law of natural selection governing the creatures on earth and to become superorganisms. If we can control the aging process, change the limits of life, and maintain our youth for a long time, it can be said that the long-held dream of humankind has finally been realized. This dream, however, may come true at a price as it will challenge the foundation of the human community that has been created based on the biological limitations and will push us to look for an entirely new kind of community.

Death is the end of an individual life that all humans will have to face someday. The reason that a war or a communicable disease epidemic poses a grave threat to humans is that it will bring about death. The fear of death has driven humankind's advancement, and whenever such horror grew exponentially in the past, humankind made a giant leap forward in philosophy, science, and medicine. As death is an absolute fear that is difficult to share with others, individuals with an intrinsic desire to carry on with life, when faced death, often change their attitude and improve their lives to avoid the sudden interruption that death brings and try to make their lives somehow connected. Therefore, individuals at the brink of death are likely to reflect on the life goals they pursued in the past and, in some cases, embrace a new attitude toward life, placing more significance on their family and friends than on worldly values like power, money, and honor. When judged from the perspective of the human mission to hand over one's own property (not necessarily only genes) to one's descendants, this attitude can be said to stem from a desire to invest in something concrete and certain, such as one's memories of the people closer to him/her.

Death, however, occurs not only in an individual level but also in the level of the cells in the human body. A cell dies in either of two ways. First, it becomes necrotic or ruptured and finally dies when it becomes difficult to survive as its environment becomes worse, such as when it is attacked by toxic substances or infected by a virus. There is another type of death, however, which allows the cells constituting an organism to kill some of them to maintain their overall function when some cells start to deteriorate in function. This is called "apoptosis." The former type of cell death can be considered death by homicide, whereas the latter can be considered death by suicide.

For example, cells exposed to radiation are damaged at macromolecules such as DNA or proteins in the cells, resulting in the loss of cellular function. Although these cells can still function to some extent, it will be much more advantageous to the human body to eliminate them or to replace them with new cells than to allow them to continue to function in a limited range when viewed in terms of overall function of the human body. This is the reason why the less-functioning cells initiate their own apoptosis program to kill themselves. If this program is not available, cell regeneration, in which the old cells are constantly replaced by new cells, will not occur, triggering a rapid decline in the function of the organs constituting the human body, thereby lowering the survival ability of individuals.[112]

Apoptosis also blocks inflammation in advance. When the cells in a tissue lose their function after being attacked by external factors, the inflammatory cells defending the body from outside attacks are mobilized. The inflammation inevitably spreads to the surrounding tissues in this case, affecting even the normal-functioning cells, thereby deteriorating the function of all the tissues. If the cells under attack, however, are eliminated by inducing a suicidal reaction via apoptosis, the tissue can maintain its normal function as inflammation is not induced. To conclude, apoptosis can be deemed to be a device designed to maintain the survival ability of humans through the prevention of tissue damage by triggering the suicidal reaction of the cells.

In apoptosis, it is not clear if the cell commits suicide of its own accord or if a higher control tower is ordering it to kill itself. As alliances, however, had been firmly established among the cells when they evolved from single cells to multicellular organisms, it will not matter much whether the cells commit suicide of their own accord or a higher control tower is ordering them to kill themselves when the sacrifice of some cells is beneficial to the tissue or the individual organism. Again to say, if death at the cellular level is beneficial to the whole body, then the death of the cell is chosen by the apoptosis program, and the life of the individual persists through the death of the cell.

It is not clear exactly where this death device came from, but it is highly probable that the device may have been left by microorganism that had evolved into mitochondria when the eukaryotic cells were born as a result of the symbiotic unity of heterologous microorganisms. The mitochondria play the role of keeping a healthy environment in the cell through division and fusion reactions. A close look at the progression of apoptosis will show that the division and fusion of the mitochondria are not sustained in harmony as division occurs more frequently, forcing the mitochondria to become fragmented. This is how apoptosis unfolds. What is interesting here is that the function of maintaining the vitality of the entire organism, such as the microscopic death device called "apoptosis," began with the creation of a symbiotic system called "mitochondria." Mitochondria are subcellular organelles that play a major role in the evolution and development of life after the emergence of the eukaryotic cells. The fact that a death device, such as apoptosis, also begins in the mitochondria signifies that death plays an important role in the evolution and development of life. In other words, the apoptosis of cells is a program designed to maintain and promote the fundamental vitality of humans, suggesting that the death of individual humans may also be a necessary program to maintain the vitality of the entire human race.

Aging is a natural phenomenon

The questions of "what does constitute the world?" and "what is life?" may have been constantly raised ever since humankind became aware of its existence. Also, "what is death?" has perhaps been as important a question as "what is life?" because death is inevitable and signifies the end of life. On the other hand, aging can be thought of as the halfway point between life and death. As aging itself, however, was not a common phenomenon in the past because the average lifespan of the humans then was relatively short, people would have paid much less attention to it. Then, is aging simply a process leading to death, or is it an abnormal health condition?

Aging is characterized by the signs or symptoms, such as being less responsive to external stimuli, a fragile physiological balance, and vulnerability to disease. Aging has been perceived as an inevitable phenomenon leading to death, but the expression of aging does not occur simultaneously across all organs at any particular moment. Wrinkles begin to develop when people are over 30 years old, cognitive abilities begin to fall in the late 30s, and farsightedness occurs in the late 40s, making it difficult for the person to see things that are near. After people pass the age of 60, degenerative changes begin to appear in the joints, and after 70, hearing loss begins. The incidences of chronic diseases and late chronic diseases also rise significantly with aging, especially neurodegenerative diseases like Alzheimer and Parkinson diseases.

As people, when still young, have abundant cells in the muscles, heart, and brain (they are mainly composed of postmitotic cells that are not replaced by new cells), the apoptosis, which kills the less-functioning cells, is activated to maintain the functions of said organs just like other organs. However, the number of cells in these organs is thus reduced with aging, because the cells that are killed are not replaced by new cells. Ultimately, because the organs require a certain volume of cells for their function, it is difficult for them to reduce the number of cells any further, thereby hindering the normal operation of apoptosis. Further, inhibition of apoptosis occurs not only in the postmitotic cells but also in the mitotic cells as people begin to age. In other words, the ability of apoptosis itself to remove damaged or dysfunctional cells is reduced as people age, impairing the functions of all the organs in the body.

Meanwhile, the cells have a self-phagocytosis function called "autophagy," which helps them recover their function by cleaning or regenerating the damaged molecules or organelles within them when they are damaged. However, the autophagy activity also deteriorates as people age. As a result, the aged cells of the organs made up of postmitotic cells, like the muscles, heart, and brain, lose their function gradually because they can hardly recover from damage by autophagy or by apoptosis, which again leads to functional deterioration at the level of the tissue. As the cells with a deteriorating function remain and do not die, the decline in the functions of the muscles, heart, and brain, whose cells have already been reduced in number, progresses more rapidly. As the amount of muscle is reduced and the cardiac and brain functions fall, causing debilitation, people eventually face death.

Looking into the process of cell damage in details, we can observe that cell damage is mostly caused by glycosylation, which is the accumulation of intracellular saccharides that damage the other molecules and organelles, as well as by oxidative stress, in which the reactive oxygen species attack the intracellular structures. In particular, such an attack causes a decrease in the number of mitochondria and their function. As the progress of aging hinders recovery from such loss and functional deterioration, the mitochondria become increasingly unable to perform their normal function.

On the other hand, the mitochondrial dysfunction is also known to play a central role in chronic diseases. If oxidative stress occurs in the mitochondria, lowering their function, it might be perceived as a danger signal for an external attack, and an inflammation reaction could be triggered. If chronic oxidative stress occurs through lifestyle factors such as an unhealthy diet, smoking, or chemical exposures, the inflammation, although low grade, is constantly caused, and then chronic diseases like diabetes mellitus, cardiovascular disease, and cancer also occur. The chronic diseases themselves are hindering normal function of humans, accelerating the aging process more.

Ultimately, the most obvious way to prevent aging, cancer, and other chronic diseases is to maintain the function of the mitochondria. To do this, it is important to reduce one's caloric intake and reduce one's exposure to external hazards, such as chemicals. If the glycosylation or oxidative stress is reduced, an inflammation reaction is not likely to occur, and the mitochondria will grow in number again and become sufficiently activated to prevent premature aging or chronic disease.

On the other hand, however, we should understand that aging is a natural phenomenon that occurs along the lifecycle from birth to death, although the premature aging process needs to be prevented. As life changes from birth to growth, development, and degeneration, the process of aging naturally leads to death, which, we can say, is the natural law. Therefore, although it is possible to slow down the rate of aging, it may not be possible to stop it completely or to reverse it, as far as we are biological humans.

The paradox of eternal life

In a biological sense, "eternal life" can be defined as a state in which an organism does not succumb to death even after a time because it does not undergo the process of aging. For example, the cells of planarians do not have a limited lifespan because their old cells are constantly replaced by new cells generated from their stem cells. Although they can be caught and eaten by other predators or can die because of changed habitats, there is no limit to their biological lifespan in a safe environment. There are two types of planarians, however: one that maintains its genes through asexual reproduction and another that preserves its genes through sexual reproduction. Interestingly, there is no limit to the lifespan of planarians produced via asexual reproduction, whereas the lifespan is limited to 3 years in the case of those produced via sexual reproduction. In other words, when genetic information is exchanged between two opposite sexes to give birth to offspring, the lifespan is limited.[113]

As such, planarians wonderfully show the relationship between gene delivery and lifespan. As is clearly shown, if genes are transferred to the offspring via sexual reproduction, the lifespan shall be limited to the constant length, and if an individual organism can maintain its genes for a long time or transfer to the offspring via asexual reproduction, its lifespan can be extended. In other words, gene delivery and lifespan are closely related. In the case of the former, although the lifespan of an individual organism is short, excellent genes more suitable for the environment may appear through a mix of genes in the process of sexual reproduction, and the possibility of genetic advancement is much higher at the species level. On the other hand, if one gives birth to offspring with the same genes via asexual reproduction, there could be no limit to the lifespan of the individual life, but individual organisms may end up living only under limited environmental conditions because there is no way to produce excellent genes that are better suited to different environment at the species level.

In fact, the very existence of human lifespan means that we have a mechanism to produce genes that are more suitable for a given environment along the line of succeeding generations. Thus, if lifespan ceases to exist and humans are immortal, these mechanisms will disappear as well. The selection of a mate in the mating process, which is included in sexual reproduction, is in fact a result of competition for securing superior genes. Therefore, if the limitation of the lifespan disappears, the competition for securing superior genes and mate selection will disappear as well. In fact, the driving force behind the history of most organisms, especially human history, was the competition to secure superior genes. What happens to humans then if this competition disappears? Perhaps when an immortal human being is born, the competition for securing superior genes will cease, and humankind may lose the most important purpose of its life. In other words, vigorous human activities such as sex, competition, and love may no longer exist.

To conclude, at the moment humankind acquires everlasting life, any further advancement of human life will be halted as well because eternal life can no longer evolve further after it loses the motivation to compete for natural selection. This is a loss of vitality inherent in life, and eternal life will fall into the paradox of loss of vitality. Therefore, when we increase the human lifespan dramatically by using various human enhancement devices or regeneration technologies, it is not known at present how the civilization will unfold. Civilization, however, must be completely different in nature from what we have seen in human history where civilization has been achieved by overcoming adversity and through suffering from competition for natural selection.

Death, a prerequisite for life

The most important organs that determine the human lifespan are the brain and the heart. The cells of these two organs are not replaced by new cells when they reach the end of their life because they are mostly made up of postmitotic cells. Therefore, the brain and heart will no longer be able to function after a certain warranty time, eventually leading to death. With the advancement of new technologies, however, like the supply of new cells by stem cells and the insertion of artificial organs, the brain and the heart are expected to be able to function well beyond their respective biological lifespans. Likewise, organs like the liver, kidneys, and pancreas may also be able to function beyond the limits of their respective lifespans. The life expectancy of humans may therefore well exceed 120 years, which is currently considered a biological limit.

Even if the human lifespan can be extended, however, the human bodily function will generally deteriorate over time unless most of the human biological organs are replaced by better mechanical organs. For example, the blood vessels age with time and become hardened, and the endothelium becomes rough. Even if we supply new cells using endothelial stem cells or regularly clean up the small vessels with small robots (it will be possible in the future!), it will be difficult to stop all aging processes of blood vessels. Moreover, the aging vascular systems cannot completely be converted to mechanical systems because the blood vessels are not organs that function independently but are connected to all the organs of the human body. As such, even if all the organs in the human body are maintained well, aging cannot be completely prevented, and humans are bound to have a lifespan anyway.

Ultimately, individual human immortality cannot be realized as long as we exist as *H. sapiens*. The singularity suggested by the futurist Ray Kurzweil signifies the point at which technological progress can be achieved at an exponential pace and that the power of AI can develop into an unimaginable level such that we will be able to solve nearly all the technical problems that we cannot solve at present. It suggests that humans will have advanced beyond the limit of death to attain immortality at a certain point. However, this claim was not reasonably inferred because biological individuals cannot overcome death as far as they are humans. Therefore, death will inevitably exist unless all humans' biological organs are replaced by machines, which actually means losing their biological human characteristics.[114] Therefore, death of biological human will still exist even if humans can live for a considerable period of time beyond the limits of their biological lifespan. Paradoxically, humankind is hard to sustain without the death of an individual human being, because people will be motivated to reproduce only with the presence of death. In fact, death is the door to a new life, which will ensure the continuation of humanity well into the next generation. In a nutshell, death is a precondition for life.

As the mortality rate is expected to fall to a very low level by the end of the 21st century, natural death will not occur that easily, and the proportion of the elderly people in the population will explode with the increase in the human lifespan. At the same time, the number of newborn babies will be very small because birth rates will be greatly reduced; therefore, the majority of the population will be composed of elderly people. If the population is not reproduced appropriately and humanity loses its motivation for further development, the human society will be stagnant and a crisis will occur in the sustainability of humanity. Therefore, it may be desirable (of course, very cautiously!) to maintain the share of the elderly people in the population below certain levels, which may necessitate the control of the demographic structure by restraining artificial life extension beyond the biological limit of human life. Perhaps we may need a program to remove the human body strengthening device and life support and to die with dignity at some point to achieve an ideal society where humankind maintains vitality to ensure sustainability.

5.4 For the Sustainability of the Human Community

The history of human biological evolution will draw to an end if it becomes possible to manipulate genes so that people will no longer get sick or it makes them even better equipped. As humanity is approaching a new state both physically and mentally due to technological enhancements, it is difficult to accurately predict the problems that will arise in the future. Further, these changes will not transpire at the current rate but at an exponential rate. Various regions on earth or even in universe will be explored and used as settlements in the future, and new diseases will inevitably occur from the new environment. Therefore, the future will give us a challenge if we pursue continuous growth at a rapid pace without preparation. Now is the time to put more emphasis on fixing the damaged parts of the human community.

Biological evolution is over

The biological evolution from the first living creature to the *H. sapiens* was a remarkable accomplishment, but it took a whopping 3.5 billion years to happen. The long evolutionary journey began when the genes contained in the first creature cloned themselves and produced descendants. It was also the time when one of the descendant genes adapted better to the environment, survived the other competing genes, and formed another descendant, and when this process was repeated over and over again. A myriad of such repetitive processes came up with slightly different descendants rather than perfect clones, thereby creating a biological world that is full of diversity. One of such descendants is the *H. sapiens*, the human race of today.

The *H. sapiens* dominated or exterminated other creatures using fire, tools, and language and began to live scattered throughout the four corners of the globe for 50,000 years after migration out of Africa. Those who invented needles and wore clothing could endure the cold weather of the Eurasian continent. They have since made countless tools contributing to the development of civilization. Today, there are many more items that one can attach to his/her body, such as glasses, watches, and shoes, along with much more developed apparel. These tools have become almost natural parts of the body, and will keep evolving even more in the near future. Eyeglasses, now even more developed, have become a medium for accepting and analyzing outside information in real time, and watches have become tools that connect with all neighboring media to make our lives more convenient, while shoes and personal transporters will be further advanced as well to help us move more easily and conveniently.

Such personal equipment made our lives more convenient as they were designed to be attached to our body and enhance our capability. In the next phase of human development, there will be an increasing number of people who have artificial devices implanted in their body to enhance their mental and physical abilities. Those with such implants can be physically more powerful, run faster, boast higher intelligence, and enjoy more advantages in all aspects of life than those without them. In fact, the people who will have ended up having enhanced human capability thanks to such implants can be said to become superorganisms beyond the capacity of biological entities. Their biological and mechanical elements for the mental and physical abilities will gradually work together in such a way that will make it almost impossible to distinguish them.

On the other hand, the power of biological evolution based on natural selection has gradually weakened ever since humanity entered the period of civilization. Especially as the biological evolution that is purely based on genetic variation has taken much longer than the rate of social development of late, it is now difficult to detect the influence of biological evolution in the human lifespan, disease, and physical characteristics, all of which had been deeply influenced by the biological conditions in the past. Rather, the current artificial environment created by humans has a greater impact on their biological characteristics. Moreover, the history of human biological evolution will draw to an end if it becomes possible to manipulate the genes that cause disease or inferior abilities so that people will no longer get sick or it makes them even better equipped.

Exponential speed of change

Thanks to the language beyond simple vocalization or means of short communication, the *H. sapiens* were able to carry out difficult collaborative tasks, which led them to dominate all other species. Despite its relatively inferior physical abilities, humanity succeeded in becoming the ruler of the planet by means of language because language-based collaboration was so efficient. Such great tools of language, however, are very slow and inaccurate compared with the Internet, the dominant means of communication today. Accordingly, the future human communication devices will go beyond the limits of language between persons. In other words, the physical and mental abilities of individual humans will be improved not only by the use of human augmentation devices for communication but also by the speed and accuracy of cooperative works facilitated by communication among the human–human and human–machine networks.

Computerized or mechanized human enhancement devices are created by imitating the human body structure, and their functions are designed by emulating the functions of diverse organs constituting the human body. As biological evolution proceeds in ways that will change the structure and function of living organisms to one that is more appropriate to the given environment, the structure and functions of the human enhancement devices are not completely different from those of humans. Rather, they can be viewed as devices that change the human structure and functions to make them better fit the environment of the future society. In a sense, although the development of human enhancement devices is not a biological evolution, it is another form of evolution based on the biological structure and function of humans.

The ability that will be enhanced most prominently among the many abilities of the mechanized human being will be intellectual ability. As computerized devices are directly or indirectly connected to the human brain via the interface, humans will be equipped with a very fast information processing capability along with a tremendous amount of knowledge. Given the current speed of development of computer technology, this is not a far-fetched prediction. When these devices are all connected to the brain, however, the load that the brain must bear will also become considerable. Thus, the state where the biological and mechanical parts coexist leads to a competition between the two kinds of parts. Perhaps because the speed of mechanical development is much faster than the rate of biological adaptive development, the mechanical part can take on the role of the biological part considerably, and the biological part can be degraded rather than further advanced.

For example, we know that using a smartphone scheduling program is much easier and more accurate than managing schedule manually depending on one's memory. Therefore, it is now possible to manage one's schedule accurately without memorizing it, but on the other hand, the ability to memorize one's schedule has deteriorated. In other words, if programs like memory, information processing, and judgment are interlinked with the brain via an interface, the biological counterpart that plays the corresponding roles is likely to be degraded rapidly. As the biological program of the brain controls not only the intellectual ability but also the physiological action of the human body, such functional deterioration can pose a serious challenge to human self-survival.

Ultimately, as humanity is approaching a new state both physically and mentally due to technological enhancements, it is difficult to accurately predict the problems that will arise in the future. It has been consistently confirmed in the history of human disease, however, that the incidence of disease increased when people were exposed to a new environment until the corresponding adaptation was realized, and such cases were sometimes serious enough to change the overall course of history. In particular, the future enhancement of human capacity will be occurring continuously, rather than being a one-time event. It is also expected that the rate of change will be much faster than the rate of human adaptation, making the resulting impact serious enough to affect the destiny of humankind, not to mention the human illness and health. Further, these changes will not transpire at the current rate but at an exponential rate in the future. As the changes that will take place over the next several decades may be equal to the changes that occurred in the past several hundred years as measured by the present level of technology, we may experience tremendous changes in civilization before the turn of the century.[114]

From the holocene to the anthropocene epoch

Humans are among the many species that depend on the global environment. Therefore, if the shield of the global environment is shattered and broken, humanity will lose the basis of its existence. The geological, physicochemical, and biological elements of the earth's environment, however, and the ecological environment associated with them, are being destroyed or transformed at an unprecedented rate in the history of the earth, suggesting that the continuity of human existence is uncertain. Humans are now the greatest factor in creating such changes, but they will also be among those who will experience the impact of such changes as they are part of the ecological environment.

People's habitats will keep expanding up to Siberia, northern Canada, and Iceland's frozen lands if climate change remains unabated and the global temperature continues to rise by 3–4 degrees on average above the surface temperature of the present era. Looking back, the reckless development of the African or South American rainforests led to the emergence of new infectious diseases as the viruses that had taken the animals in the rainforests as hosts frequently came into contact with humans. Therefore, climate change not only causes such natural disasters as receding glaciers, floods, droughts, desertification, and sea level rise, all of which, combined, could set off a drastic change in the ecosystem at a global level, but also it can be an opportunity for the viruses and bacteria that are currently dormant in the frozen ground or glaciers to emerge as new risk factors threatening human health.

In addition to climate change, the global environment is facing a variety of crises, such as the pollution of the atmosphere, soil, and water; the excessive cultivation for crop production; the disturbance of the biological environment due to livestock breeding and fish culture; and the deterioration of the natural environment due to the unbridled land development. Evidences show that climate change is threatening the existence of Pacific islands by raising the sea levels, that the rainforest development in Africa leads to the occurrence of new infectious diseases (e.g., Ebola), and that heating with coal in the winter season has caused widespread air pollution in East Asia and other vexing environmental problems. The fundamental driving force of such change is the increase in the population and in the per-capita consumption, which has grown more rapidly recently than in the past. As the population increases, more food is needed to feed people, and more space is needed to accommodate them. Accordingly, if this trend continues, the earth, which has already been overused, may face an irrevocable crisis within this century in the geological, physicochemical, and biological aspects.

The rapidly progressive decline of species diversity is expected to reduce the existing species by 30% in the next few decades.[115] The scale of devastation will be similar to the mass extinctions of biological species in the past history of the earth. This change in the global environment suggests that the Holocene epoch, during which biological species became diverse in warm climate with the end of the last ice age, and humankind greeted the advent of civilization, is drawing to an end. This means that humankind should live under the conditions created by the new global environment. The fundamental difference between the current changes and the global environmental changes recorded in the past history of the earth, however, is that the current planetary changes are not caused by nature but by humans. In other words, the Anthropocene epoch, in which humanity exert a decisive influence on the environment, has already started.

New environment, new diseases

Mankind has been constantly exploring and pioneering new settlements. It is this pioneering spirit that made our ancestors venture out of Africa some 50,000 years ago, and spread to the four corners of the world to ultimately build civilizations. After humans settled around the world, they once again migrated and exchanged goods by developing trade route such as the Silk Road or sea routes. Today, nearly 10% of all people move every year for travel or business purposes. Perhaps the convenience of mobility will increase even more in the future, and the world will look almost like one global village.

In the future, various regions on earth will be explored and used as settlements, an act aimed at avoiding the competition in an existing area or at obtaining the new resources that are necessary for living so as to gain greater profit, or at avoiding the significant risk lurking in the current settlement. The act of pioneering this new settlement is very likely to extend beyond the planet Earth and well into other planets or satellites because living on earth may itself expose us to irreparable catastrophic threats, such as a comet collision or a nuclear explosion, and in some cases, it is easy to get the resources from space that are unavailable on earth.

In this regard, we can learn from the history of our ancestors who explored new areas and expanded human settlements. In the 15th century, when a group of people sailed out to sea for an extended voyage, they suffered from symptoms that were not seen before. A disease with symptoms like bleeding gums, falling teeth, weight loss, and painful bones and muscles, and eventual death, spread among the crew. In the 18th century, James Lind claimed that the disease was caused by the lack of a specific nutrient found in oranges and lemons, which later became known as vitamin C. As it turned out, the disease was scurvy, which occurred as the crew was unable to ingest fresh vegetables or fruits as a result of their long marine life in the Age of Exploration.

If we venture into a completely new environment like the universe, new diseases will inevitably occur because we will be in an environment that is very different from the earth's environment. A study of astronauts who were out of the earth's orbit for a long time showed that their cardiovascular mortality rates went up four- to fivefold those of people who did not leave the earth's orbit.[116] It is presumed that new environmental factors, such as various cosmic radiations and zero gravity, which could not be experienced in the global environment, were the probable causes. But we don't know further. The process of building a space residence or of creating a new settlement in the universe will not be easy as our ancestors experienced many failures and faced many challenges in the Age of Exploration, which began in the 15th century, or during the expansion out of Africa some 50,000 years ago. Mankind may be able to successfully develop settlements by adopting human enhancements, using robots that support humans, or developing unparalleled technological systems. The new diseases emerging in this process, however, will make humanity face great ordeals and challenges, as has been the case in the past.

An organism, the human community

The expansion of the universe may not occur in a conceptual domain completely detached from the survival of humankind. Perhaps it is our fundamental environment that defines our very existence. The presence of such expansion can be explained as an act of energy or as a phenomenon that occurs when energy is released explosively. Newton's law of gravity or even Einstein's principle of general relativity, which attempted to explain Newton's theory with the concept of time and space, may not be an eternal truth but a tentative law of physics that can be observed only during a time when the universe are still expanding. For example, if you ignite a firecracker, it spreads in circles and beautifully decorates the night sky, but once the energy dissipated, it stops spreading and falls to the ground. Perhaps the phenomenon that we are currently observing is like observing the moment of such spread over the night sky because the laws of physics, such as the laws of gravity or general relativity, can also vary along with the stage of universe expansion. In other words, the physical principles that we currently know are not eternal rules, and the biological phenomena that we have observed in the history of our planet or *H. sapiens* can be even more tentative phenomena.

Mankind will probably not have an opportunity to observe all biological phenomena, which is changing over time, by experiencing both the expansion and extinction stages of the universe. It is also uncertain if the next hominid species will continue even if humankind will succeed in overcoming numerous adversities in the coming years and hands over all the experiences and heritage of humankind to the new species after the demise of the modern humans. Furthermore, it is even more difficult to predict if our future descendants will look at the modern humans with respect and gratitude or as just one among their primitive predecessors. We are not sure for now if humanity's brilliant achievements, such as the control of chronic diseases, the strengthening of human capacities, and the extension of human life beyond its biological limits that are expected to occur in the coming decades or at the end of this century, will be a blessing to humankind in the long run or will turn out to be a serious challenge to overcome. We can see, however, that the decisions of humankind at present, which will open up the future, can change the future direction of development. Therefore, we should make the best decision to avoid a catastrophe and to prepare for a better future.

In the coming years, the human lifespan will increase along with the substantial change in the population structure, which will again change the residential, dietary, and other living environment, which will also affect the entire ecosystem. These changes can again affect the relationship of coexistence in the planetary ecosystem and lead to the occurrence of new diseases if adaptation to such environmental change is not sufficiently realized. The future will also bring new physical, chemical, and sometimes biological exposure, which may cause an epidemic of a disease that has never been experienced by humankind. In other words, even if chronic diseases can be successfully controlled, it may not be possible to entirely prevent the occurrence of future diseases.

What is more problematic is that if technology develops at a very high but uneven speed among population groups, the gap between the group that can fully enjoy the technology and the group that cannot access it easily will keep widening. Moreover, the difference between these two groups may not be just about opportunities to enjoy cultural life presented to them; it can escalate and come to threaten the homogeneity of the *H. sapiens* because it can lead to a difference in the physical and mental abilities as well as in the disease pattern and lifespan. The new environment, the diseases caused by it, and the strategies to overcome it have so far transpired within the scope or biological capabilities of the *H. sapiens*, but if there will be challenges beyond these, and if the matching strategy is also beyond the scope and capabilities of the *H. sapiens*, the fundamental question on human sustainability will be raised.

When people's lives are in crisis, it is advantageous for them to use their energy resources to maintain their survival, primarily by repairing the damaged parts inside rather than using the energy for growth and reproduction, because survival takes precedence over growth and reproduction. The human community can also be considered as a living organism. Therefore, if we pursue continuous growth at a rapid pace without repairing the damaged parts of the human community, the community can lose its power to keep sustainability. Now is the time to put more emphasis on fixing the damaged parts of the basic elements that make up the community, such as deteriorating ecosystem, widening disparity of living conditions, and the inequality in terms of science and health benefit, because a healthy future society can be created only by securing the sustainability of humankind.

Epilogue

1. For Conquering Diseases

The era in which disease has had a profound impact on human life, death, and longevity is now drawing to a close. At the same time, medical science, which has developed with a focus on diagnosing and treating diseases, faces the challenge of adapting to the rapid changes that are coming. This is mainly because not only have changes in disease patterns, increased life expectancy, and an increase in the elderly population changed the disease itself, but also the changes in the disease have had a profound impact on human life. In addition, disease management methods are changing as our understanding of the effects of diverse lifestyle and environmental factors and genes on the occurrence and progression of disease deepens. In fact, we are entering an era in which we should focus on "health" rather than on "disease." Therefore, rather than simply diagnosing and treating disease, the methodological exploration and practice of identifying all of the factors affecting disease development and progress and managing the health by changing the parts that need to be corrected will be the core of medical care.

The most important tasks, in particular, for the successful management of diseases are to develop medical technology and to change the medical system so that anyone can gain access to medical care and easily treat disease and maintain his/her physical, mental, and social functions. To do this, comprehensive medical care must be provided, along with the "systems medicine" approach. In other words, a medical care system should be created in which diagnosis, treatment, and management are performed seamlessly and continuously from the home, school, or workplace to the hospital. A comprehensive and in-depth assessment of the health status of the community members should be made at all times, and a medical system should be established for analyzing both biological and clinical data as well as other risk factors to which people are exposed in their daily life. In addition, it will be necessary to analyze such complicated information and to communicate with people efficiently, and to help them manage their own health in an easily accessible way. This is because a useful recommendation cannot be carried out unless it is easy to understand and is practical.

Along with the efforts to create such medical care system, the concepts of disease and medical education should be changed as well. The treatment should be centered on individual patients or persons rather than on disease. If this change based on systems medicine is successfully realized, we may have a chance to create a healthy society. Moreover, as a result of these changes, we may be able to see the end of the Age of Disease in the future by gaining the ability to prevent and cure diseases to a considerable extent, including infectious, chronic, and even late chronic diseases.

2. Let's Prepare for the Next Chapter of History

The healthy future society that we are dreaming of is a sustainable society, with each individual enjoying a happy life as a member of the community. In addition, the sustainability of humankind means that descendants are born continuously, and humankind as *Homo sapiens* persists and flourishes. In addition, the happy life of humans means that one enjoys material abundance and maintains vitality not just as an individual human but also as a member of a community. A healthy and wonderful future, however, will not come easily. To achieve the genuine control of disease and a healthy society, there are challenges that must be overcome, such as the resolution of inequality, controlling the rate of change, and the dissolution of the conflicting interests of individuals and communities. Particularly if medical technologies like life extension or human enhancement beyond the limits of life are used indiscriminately, fundamental problems related to life, death, and human continuity may arise, confusing us seriously.

In fact, the human desire to take a more favorable position in the competition for survival is bound to cause change. Thus, we cannot stop the efforts to profit from the advancement of science and medical technology. Moreover, preventing such change can be likened to prohibiting efforts to create a better future. The key issues here are to control the pace of change and to coordinate with one another regarding the change. To secure the sustainability of humankind, it is necessary for us to adjust the pace of change so that we can have enough time to adapt to such change without leaving anyone behind.

If we look closely at the rapidly changing science and technology and living environment, we will see that it is not the entire humanity that is taking the lead to make such change happen. If we look at the human communities on earth, we will see countries with varying levels of economic development: the advanced countries are leading the economic growth, and the underdeveloped communities are incapable of making economic growth happen and are suffering the damage caused by changes made by the developed communities. The fact that the rate of change is rapid suggests that this spectrum is widening and the inequality is growing. The inequality within the human race can trigger a crisis that can threaten the sustainability of humankind, thereby potentially leading to a local conflict or a destructive ending to another world war. What is more frightening is that the present socioeconomic inequalities may develop into biological inequalities, which may end human history as we know it. The possible destruction or end of human history indicates that *H. sapiens* may not be facing a brighter future.

We are standing at a crossroads: we will either conquer disease assisted by positive changes like the development of science and technology and lifestyle changes, or we will face a devastating end to human civilization due to the deepening inequality among humans, particularly the gap between the rich and the poor, and the unbridled use of science and medical technology. There is a new challenge ahead of us that we have not faced in the past, and if we do not address it properly, we may not be able to control disease, or we may not have a wonderful future to pass on to our descendants. In this regard, we need to prepare for the next chapter of human history by urgently addressing the threats we are facing in the present age.

3. Finishing This Book

In this book, I sought to explain methodological solutions for the conquest of disease, which has been a long-time dream of humankind. I also intended to explain the various problems that may arise in the course of conquering diseases. And I explained that even though most of the diseases that we are currently experiencing disappear in the future, it is highly likely that new diseases will appear. I argue as well that death is a necessary device to maintain human life.

The problems that have been raised in front of humanity, such as hunger, war, infectious disease epidemics, and chronic diseases, which humankind has experienced since the age of hunter-gatherers, are never easy challenges. There will be more challenges in the future because the problems we will face may threaten the very existence of humankind. I carefully anticipate, however, that having acquired much experience while going through many hardships and adversities in the past, humankind may overcome these problems with wisdom. The future of civilization will be much better, therefore, but only if we address these problems properly.

References

1. Hong Y-C. *The origin of diseases*. Nova Science Publishers; 2015.
2. Stearns SC, Jacob CK. *Evolution in health and disease*. 2nd ed. Oxford University Press; 2007.
3. Edmond Dounias, Alain Froment. When forest-based hunter-gatherers become sedentary: consequences for diet and health. http://www.fao.org/docrep/009/a0789e/a0789e07.html.
4. Cohen MN. *Health and the rise of civilization*. Yale University Press; 1989.
5. Robson B, Baek OK. *The engines of Hippocrates: from the dawn of medicine to medical and pharmaceutical informatics*. John Wiley & Sons; 2009.
6. Guns DJ. *Germs, and steel: the fates of human societies*. W.W Norton & Company; 1999.
7. Nunn JF. *Ancient Egyptian medicine*. University of Oklahoma Press; 1996.
8. Zhenguo W, Ping C, Peiping X. *History and development of traditional Chinese medicine*. Science Press; 1999.
9. Toby Evans S, Webster DL. *Archaeology of ancient Mexico and Central America: an encyclopedia*. Garland Publishing, Inc.; 2001.
10. Martin DL, Goodman AH. Health conditions before Columbus: paleopathology of the native north Americans. *Cult Med* 2002;**176**:65–8.
11. Kurin DS. Trepanation in South-Central Peru during the early to late intermediate period (ca. AD 1000-1250). *Am J Phys Anthropol* 2013;**152**(4):484–94.
12. Black A. The "Axial Period": what was it and what does it signify? *Rev Polit* 2008;**70**:23–39.
13. In Seok Y, Ki Baek L. *Hippocrates collection*. Nanam Publishing House; 2011.
14. Kennedy MT. *A brief history of disease, science, and medicine*. Asklepiad Press; 2009.
15. Bynum W. *The history of medicine: a very short introduction*. Oxford University Press; 2008.
16. Bendick J. *Galen and the gateway to medicine*. Bethlehem Books; November 1, 2002.
17. Fears JR. The plague under Marcus Aurelius and the decline and fall of the Roman Empire.
18. Qui Chong (translated by Oh Soo Hyeon) J. *Inner Canon of the yellow emperor*. 2010.
19. History of Indian Medicine. http://quatr.us/india/science/medicine.htm.
20. Narayanaswamy V. Origin and development of ayurveda. *Ancient Sci Life* 1981;**1**(1):1–7.
21. McNeill WH. *Plagues and peoples*. Anchor Books; 1998.
22. Barnes E. *Diseases and human evolution*. University of Mexico Press; 2007.
23. Hopkins DR. *The greatest killer: smallpox in history*. University of Chicago Press; 2002.
24. Fenner F, Henderson DA, Arita I, Jezek Z, Ladnyi ID. *Smallpox and its eradication*. WHO; 1988.
25. Ligon BL. Plague: a review of its history and potential as a biological weapon. *Semin Pediatr Infect Dis* 2006;**17**(3):161–70.
26. Frith J. The history of plague – part 1. The three great pandemics. *J Mil Veterans' Health* 2012;**20**(2):11–6.
27. Kennedy M. *A brief history of disease, science, and medicine*. Asklepiad Press; 2004.
28. Lippi D, Gotuzzo E. The greatest steps towards the discovery of Vibrio cholera. *Clin Microbiol Infect* 2013:1–5.

29. Roberts JM, Westad OA. *The history of the world*. Oxford University Press; 2015.
30. Marx RB. *The origins of the modern world: a global and environmental narrative from the fifteenth to the twenty-first century*. Rowman & Littlefield Publishers; 2014.
31. Merrill RM, Davis SS, Lindsay GB, Khomitch E. Explanations for the 20th-century tuberculosis decline: how the public gets it wrong. *J Tubercul Res* 2016;**4**:111–21.
32. Brownson RC, Bright FS. Chronic disease control in public health practice: looking back and moving forward. *Publ Health Rep* May-June 2004;**119**.
33. Orho-Melander M, Klannemark M, Svensson MK, Ridderstråle M, Lindgren CM, Groop L. Variants in the calpain-10 gene predisposed to insulin resistance and elevated free fatty acid levels. *Diabetes* 2002;**51**(8):2658–64.
34. World Health Organization. *Global health observatory data*. http://www.who.int/gho/ncd/mortality_morbidity/en/.
35. World Health Organization. *Global status report on non-communicable diseases*. 2010. p. 2011.
36. Murray CJ, Lopez AD. Measuring the global burden of disease. *N Engl J Med* 2013;**369**(5):448–57.
37. Wang H, Dwyer-Lindgren L, Lofgren KT, Rajaratnam JK, Marcus JR, Levin-Rector A, Levitz CE, Lopez AD, Murray CJ. Age- and sex-specific mortality in 187 countries, 1970-2010: a systematic analysis for the Global Burden of Disease Study 2010. *Lancet* 2012;**380**(9859):2071–94.
38. Tunstall-Pedoe H, Kuulasmaa K, Mähönen M, Tolonen H, Ruokokoski E, Amouyel P. Contribution of trends in survival and coronary-event rates to changes in coronary heart disease mortality: 10-year results from 37 WHO MONICA project populations. Monitoring trends and determinants in cardiovascular disease. *Lancet* 1999;**353**(9164):1547–57.
39. Cancer trends progress report – 2011/2012 update. Bethesda, MD: National Cancer Institute, NIH, DHHS; August 2012. http://progressreport.cancer.gov.
40. Healthy People 2020 leading health indicators: progress update. USA: Department of Health & Human Services. www.healthypeople.gov.
41. Hochberg Z, Feil R, Constancia M, Fraga M, Junien C, Carel JC, Boileau P, Le Bouc Y, Deal CL, Lillycrop K, Scharfmann R, Sheppard A, Skinner M, Szyf M, Waterland RA, Waxman DJ, Whitelaw E, Ong K, Albertsson-Wikland K. Child health, developmental plasticity, and epigenetic programming. *Endocr Rev* 2011;**32**(2):159–224.
42. Mayr E. *This is biology*. Cambridge: Belknap Press of Harvard University Press; 1997.
43. Kaneko K. *Life: an introduction to complex systems biology*. Springer; 2006.
44. Lane N. *Life ascending: the ten great inventions of evolution*. W. W. Norton & Company; 2009.
45. Gauch HG. *Scientific method in practice*. Cambridge University Press; 2003.
46. Collen A. *10% Human: how your body's microbes hold the key to health and happiness*. HarperCollins Publishers; 2015.
47. Wild CP. The exposome: from concept to utility. *Int J Epidemiol* 2012;**41**:24–32.
48. Roseboom T, De Rooij S, Painter R. The Dutch famine and its long-term consequences for adult health. *Early Hum Dev* 2006;**82**(8):485–91.
49. Auffray C, Chen Z, Hood L. Systems medicine: the future of medical genomics and healthcare. *Genome Med* 2009;**1**(1):2.
50. Gilbert SF, Sapp J, Tauber AI. A symbiotic view of life: we have never been individuals. *Q Rev Biol* 2012;**87**(4):325–41.
51. Lee YK, Mazmanian SK. Have the microbiota played a critical role in the evolution of the adaptive immune system? *Science* 2010;**330**(6012):1768–73.

52. Hoppensteadt F. Predator-prey model. *Scholarpedia* 2006;**1**(10):1563.
53. Larsen N, Vogensen FK, Van den Berg FW, Nieseon DS, Andreasen AS, Pedersen BK, Al-Soud WA, Sorensen SJ, Hansen LH, Jakobsen M. The gut microbiota in human adults with type 2 diabetes mellitus differ from those in non-diabetic adults. *PLoS One* 2010;**5**:e9085.
54. Cochran G. *Henry harpending. The 10,000 year explosion: how civilization accelerated human evolution.* Basic Books; 2009.
55. Monosson E. *Evolution in a toxic world: how life responds to chemical threats.* Island Press; 2012.
56. Moi P, Chan K, Asunis I, Cao A, Kan YW. Isolation of NF-E2-related factor 2 (Nrf2), an NF-E2-like basic leucine zipper transcriptional activator that binds to the tandem NF-E2/AP1 repeat of the beta-globin locus control region. *Proc Natl Acad Sci USA* 1994;**91**(21):9926–30.
57. Liska DJ. The detoxification enzyme systems. *Altern Med Rev* 1998;**3**(3):187–98.
58. Bessems JG, Vermeulen NP. Paracetamol-(acetaminophen)-induced toxicity: molecular and biochemical mechanisms, analogues, and protective approaches. *Crit Rev Toxicol* 2001;**31**(1):55–138.
59. Conboy LA, Edshteyn I, Hilary G. Ayurveda and panchakarma: measuring the effects of a holistic health intervention. *Sci World J* 2009;**9**:272–80.
60. Ferguson LR, Philpott M. Nutrition and mutagenesis. *Annu Rev Nutr* 2008;**28**:313–29.
61. Kaufman J. Evolution and immunity. *Immunology* 2010;**130**:459–62.
62. Bravo IG, Félez-Sánchez M. Papillomaviruses: viral evolution, cancer, and evolutionary medicine. *Evol Med Public Health* 2015:eov003.
63. Coskun M. Intestinal epithelium in inflammatory bowel disease. *Front Med* 2014;**1**(24):1–5.
64. Fairweather DL. *Autoimmune disease: mechanisms. encyclopedia of life sciences.* John Wiley & Sons, Ltd.; 2007. http://www.roitt.com/elspdf/Autoimmune_Disease_Mechanisms.pdf.
65. Russell Howard Tuttle. *Human evolution.* Encyclopaedia Britannica, Inc.; 2018. https://www.britannica.com/science/human-evolution.
66. Singer W. The brain, a complex self-organizing system. *Eur Rev* 2009;**17**(2):321–9.
67. *World population ageing.* New York: United Nations; 2015. 2015.
68. Irvine GB, El-Agnaf OM, Shankar GM, Walsh DM. Protein aggregation in the brain: the molecular basis for Alzheimer's and Parkinson's diseases. *Mol Med* 2008;**14**(7–8):451–64.
69. Shaw-Smith C, Pittman AM, Willatt L, Martin H, Rickman L, Gribble S, Curley R, Cumming S, Dunn C, Kalaitzopoulos D, Porter K, Prigmore E, Krepischi-Santos AC, Varela MC, Koiffmann CP, Lees AJ, Rosenberg C, Firth HV, De Silva R, Carter NP. Microdeletion encompassing MAPT at chromosome 17q21.3 is associated with developmental delay and learning disability. *Nat Genet* 2006;**38**(9):1032–7.
70. Genetics Home Reference. *SNCA.* U.S. National Library of Medicine; 2013.
71. Fried LP, Tangen CM, Walston J, Newman AB, Hirsch C, Gottdiener J, Seeman T, Tracy R, Kop WJ, Burke G, McBurnie MA. Cardiovascular Health Study Collaborative Research Group. Frailty in older adults: evidence for a phenotype. *J Gerontol A Biol Sci Med Sci.* 2001;**56**(3):M146–56.
72. Shamliyan T, Talley KM, Ramakrishnan R, Kane RL. Association of frailty with survival: a systematic literature review. *Ageing Res Rev* 2013;**12**(2):719–36.
73. Power LN. *sex, suicide – mitochondria and the meaning of life.* Oxford University Press; 2005.
74. Shetty PK, Galeffi F, Turner DA. Cellular links between neuronal activity and energy homeostasis. *Front Pharmacol* 2012;**3**:43.
75. Seo AY, Joseph AM, Dutta D, Hwang JC, Aris JP, Leeuwenburgh C. New insights into the role of mitochondria in aging: mitochondrial dynamics and more. *J Cell Sci* 2010;**123**(Pt 15):2533–42. 1.

76. Bely AE, Nyberg KG. Evolution of animal regeneration: Re-emergence of a field. *Trends Ecol Evol* 2010;**25**(3):161–70.
77. Chakraborty C, Agoramoorthy G. Stem cells in the light of evolution. *Indian J Med Res* 2012;**135**(6):813–9.
78. Yun MH. Changes in regenerative capacity through the lifespan. *Int J Mol Sci* 2015;**16**(10):25392–432.
79. Chin L, Artandi SE, Shen Q, Tam A, Lee SL, Gottlieb GJ, Greider CW, DePinho RA. p53 deficiency rescues the adverse effects of telomere loss and cooperates with telomere dysfunction to accelerate carcinogenesis. *Cell* 1999;**97**(4):527–38.
80. Ameur A, Stewart JB, Freyer C, Hagström E, Ingman M, Larsson NG, Gyllensten U. Ultra-deep sequencing of mouse mitochondrial DNA: mutational patterns and their origins. *PLoS Genet* 2011;**7**(3):e1002028.
81. Gilligan I. Neanderthal extinction and modern human behaviour: the role of climate change and clothing. *World Archaeol* 2007;**39**(4):499–514.
82. Jinek M, Chylinski K, Fonfara I, Hauer M, Doudna JA, Charpentier E. A programmable dual-RNA-guided DNA endonuclease in adaptive bacterial immunity. *Science* 2012;**337**(6096):816–21.
83. Tinetti ME, Fried T. The end of the disease era. *Am J Med* 2004;**116**(3):179–85.
84. Carroll L. *Alice through the looking glass*.
85. Kuhn TS. *The structure of scientific revolutions*. University of Chicago Press; 2012.
86. Duffy TP. The flexner report — 100 years later. *Yale J Biol Med* 2011;**84**(3):269–76.
87. McMichael AJ. *Human frontiers, environments, and disease: past patterns, uncertain futures*. Cambridge University Press; 2001.
88. Cardis E, Krewski D, Boniol M, Drozdovitch V, Darby SC, Gilbert ES, Akiba S, Benichou J, Ferlay J, Gandini S, Hill C, Howe G, Kesminiene A, Moser M, Sanchez M, Storm H, Voisin L, Boyle P. Estimates of the cancer burden in Europe from the radioactive fallout from the Chernobyl accident. *Int J Cancer* 2006;**119**(6):1224–35.
89. Hill K, Hurtado AM, Walker RS. High adult mortality among Hiwi hunter-gatherers: implications for human evolution. *J Hum Evol* 2007;**52**:443e454.
90. Fredette J, et al. Chapter 1.10: The promise and peril of hyperconnectivity for organizations and societies. *Glob Inf Technol Rep* 2012.
91. Yeong Seong Y, et al. *The arrival of a hyper-connected society and the future of humanity*. Hanul Publishing Co; 2014.
92. Barnosky AD, Hadly EA. *Tipping point for planet earth*. HarperCollins; 2015.
93. Yu Dong K. *Civilizations in the alluvial epoch (Korean)*. Publishing House Gil; 2011.
94. David Curtis Wright. *The history of China*. Greenwood Publishing Group; 2001.
95. Friedman DB, Johnson TE. A mutation in the age-1 gene in *Caenorhabditis elegans* lengthens life and reduces the hermaphrodite fertility. *Genetics* 1988;**118**(1):75–86.
96. Kenyon C. The first long-life mutants: discovery of the insulin/IGF-1 pathway for ageing. *Philos Trans R Soc Lond B Biol Sci* 2011;**366**(1561):9–16.
97. Cogneau D, Kesztenbaum L. *Short and long-term impacts of famines: the case of the siege of Paris 1870-1871*. PSE Working Papers 2016-11. 2016.
98. Moore SC, Patel AV, Matthews CE, Berrington de Gonzalez A, Park Y, Katki HA, Linet MS, Weiderpass E, Visvanathan K, Helzlsouer KJ, Thun M, Gapstur SM, Hartge P, Lee IM. Leisure time physical activity of moderate to vigorous intensity and mortality: a large pooled cohort analysis. *PLoS Med* 2012;**9**(11):e1001335.
99. Streppel MT, Ocké MC, Boshuizen HC, Kok FJ, Kromhout D. Long-term wine consumption is related to cardiovascular mortality and life expectancy independently of moderate alcohol intake: the Zutphen Study. *J Epidemiol Community Health* 2009;**63**(7):534–40.

100. Di Castelnuovo A, Costanzo S, Bagnardi V, Donati MB, Iacoviello LMD, De Gaetano GMD. Alcohol dosing and total mortality in men and women: an updated meta-analysis of 34 prospective studies. *Arch Intern Med* 2006;**166**:2437–45.

101. Jha P, Ramasundarahettige C, Landsman V, Rostron B, Thun M, Anderson RN, McAfee T, Peto R. 21st-century hazards of smoking and benefits of cessation in the United States. *N Engl J Med* 2013;**368**:4.

102. Thun MJ, Carter BD, Feskanich D, Freedman ND, Prentice R, Lopez AD, Hartge P, Gapstur SM. 50-year trends in smoking-related mortality in the United States. *N Engl J Med* 2013;**368**:4.

103. Wilson DL. *Evolution and experiment show the way.* http://www.bio.miami.edu/dwilson/Chapt6.pdf

104. Lane N. *Life ascending: the ten great onventions of evolution.* 1st reprint edition W.W. Norton & Company; June 14, 2010.

105. Aubert G, Lansdorp PM. Telomeres and aging. *Physiol Rev* 2008;**88**(2):557–79.

106. Neill D. Life's timekeeper. *Ageing Res Rev* 2013;**12**(2):567–78.

107. Schoenhofen EA, Wyszynski DF, Andersen S, Pennington J, Young R, Terry DF, Perls TT. Characteristics of 32 supercentenarians. *J Am Geriatr Soc* 2006;**54**(8):1237–40.

108. Easterbrook G. *What happens when we all live to 100?* http://www.theatlantic.com/features/archive/2014/09/what-happens-when-we-all-live-to-100/379338/.

109. Huxley A. *Brave new world.* Reprint edition: Harper Perennial; October 18, 2006.

110. Orwell G. *1984, Harvill Secker.* June 8, 1949.

111. Dawkins R. *The selfish gene.* Oxford University Press; October 25, 1990.

112. Lu B, Chen H-D, Lu H-G. The relationship between apoptosis and aging. *Adv Biosci Biotechnol* 2012;**3**:705–11.

113. Schmidtea Mediterranea. http://www.geochembio.com/biology/organisms/planarian/#dev_stages.

114. Kurzweil R. *The singularity is near: when humans transcend biology.* Penguin; 2006.

115. Novacek MJ, Cleland EE. The current biodiversity extinction event: scenarios for mitigation and recovery. *Proc Natl Acad Sci USA* 2001;**98**(10):5466–70.

116. Delp MD, Charvat JM, Limoli CL, Globus RK, Ghosh P. Apollo lunar astronauts show higher cardiovascular disease mortality: possible deep-space radiation effects on the vascular endothelium. *Sci Rep* 2016;**6**(29901).

Index